答案之书

The Book of Answers

江婉余◎编著

哈尔滨出版社
H.P.H
HARBIN PUBLISHING HOUSE

U0666606

图书在版编目（CIP）数据

答案之书 / 江婉余编著 . —— 哈尔滨 : 哈尔滨出版社 , 2024.4

ISBN 978-7-5484-7717-4

Ⅰ . ①答… Ⅱ . ①江… Ⅲ . ①人生哲学—通俗读物 Ⅳ . ① B821-49

中国国家版本馆 CIP 数据核字 (2024) 第 039370 号

书　　名: **答案之书**
DA AN ZHI SHU

作　　者: 江婉余　编著
责任编辑: 李维娜
英文翻译: 李慧泉
封面设计: 仙　境
内文排版: 宇菲世纪

出版发行: 哈尔滨出版社（Harbin Publishing House）
社　　址: 哈尔滨市香坊区泰山路 82-9 号　　邮编: 150090
经　　销: 全国新华书店
印　　刷: 三河市华晨印务有限公司
网　　址: www.hrbcbs.com
E-mail: hrbcbs@yeah.net
编辑版权热线:（0451）87900271　87900272
销售热线:（0451）87900202　87900203

开　　本: 700mm×990mm　1/32　印张: 20.625　字数: 10 千字
版　　次: 2024 年 4 月第 1 版
印　　次: 2024 年 4 月第 1 次印刷
书　　号: ISBN 978-7-5484-7717-4
定　　价: 32.00 元

凡购本社图书发现印装错误，请与本社印制部联系调换。

服务热线:（0451）87900279

使用说明

当你为生活中纷扰繁杂的小事犹豫不决时，这本书会给你参考。

在快节奏的现代生活中，我们要面对无数选择，有些选择很容易做出决定，有些却让我们犹豫不决。

遇到这种情形，打开这本书，让足够简单的答案和暗示点醒自己。

在这里，你会找到答案。

关于问题，也关于人生。

难题几百种，最终的答案不止"是"或"否"。

让这本富含智慧与哲理的《答案之书》，治愈你的选择困难症。

本书用法——

◇把书捧在手中，闭上眼睛；

◇在心里默念要问的问题，重复一次；

◇深呼吸三次，打开任意一页；

◇睁开眼睛，获得答案；

◇按照以上步骤，寻找其他问题的答案。

打开它，

所有的问题，这里都有答案，

所有的人生困惑，都将得到指引！

再考虑一下

Think again

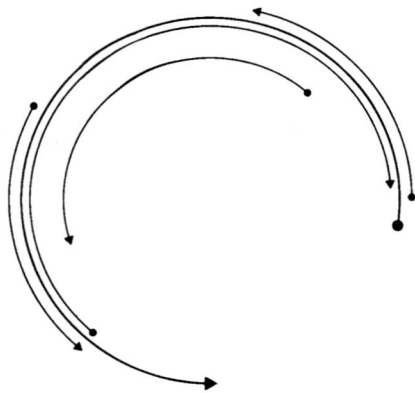

勇敢尝试

Be brave to try

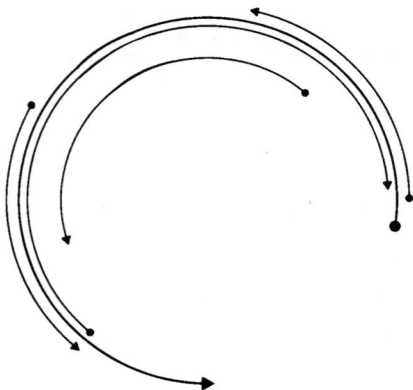

重翻一次

Open the book again

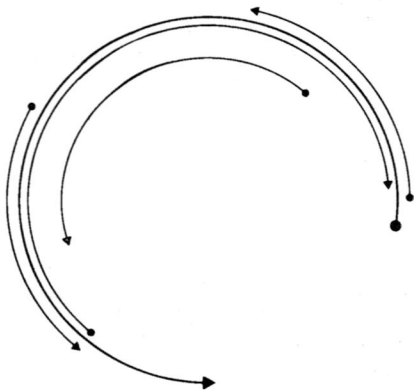

会没事的

It will be ok

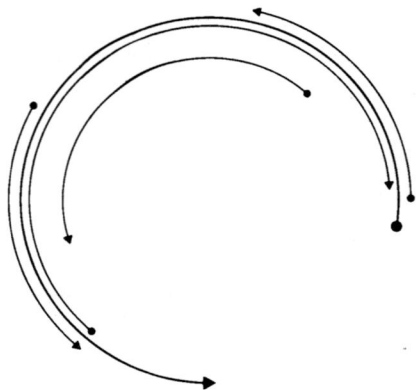

8

无所畏惧

Be fearless

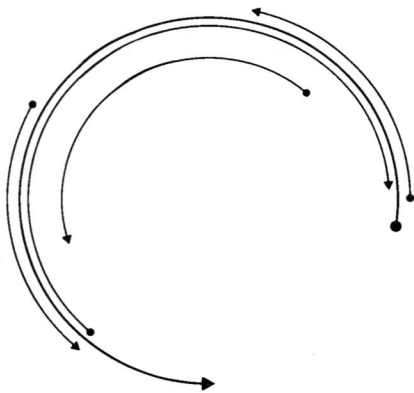

接纳多样性

Accepting diversity

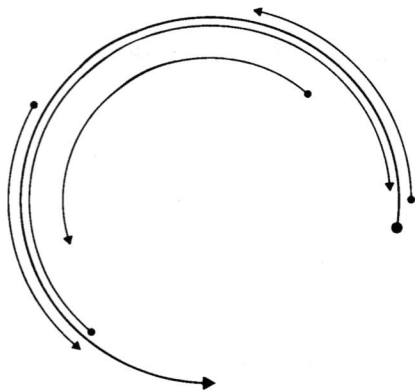

没问题

No problem

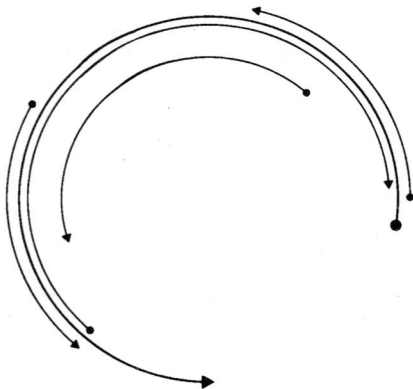

期待的总会到来

The expected always comes

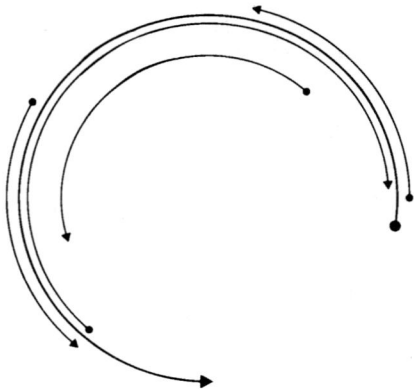

停止精神内耗

Stop mental internal friction

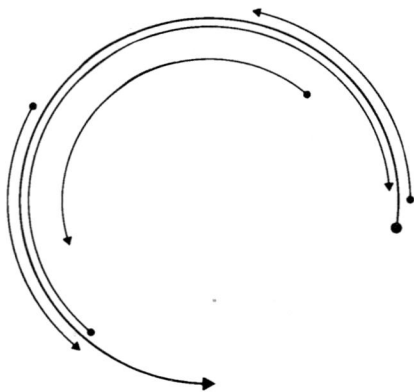

18

把握机会

Seize the opportunity

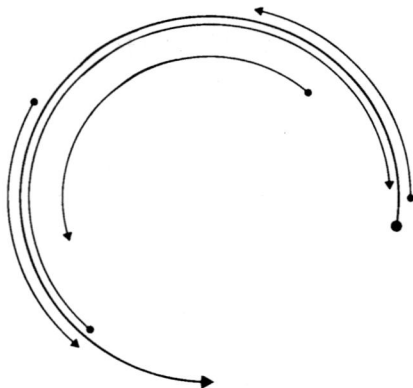

懂比爱更重要

Knowing is more important than loving

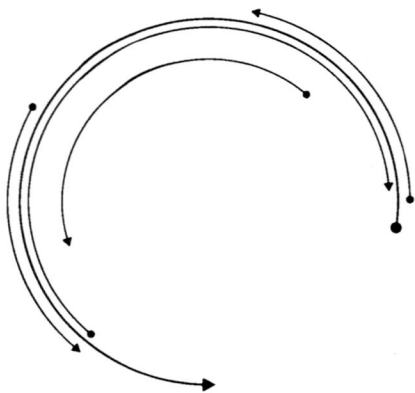

适当躺平

Appropriate to lie flat

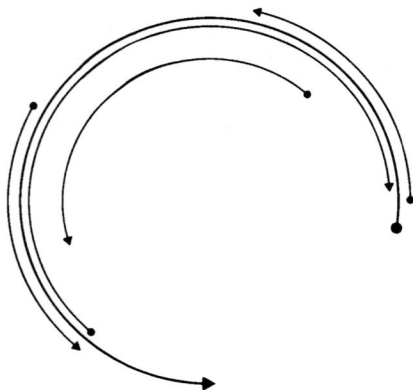

运气也是一种能力

Luck is also a kind of ability

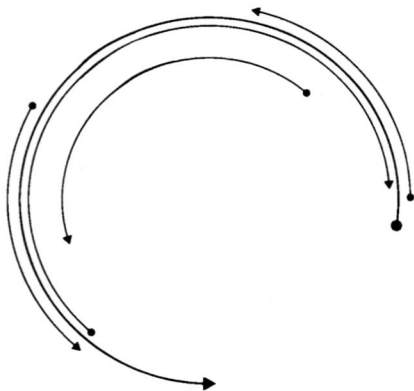

专注当下

Focus on the present moment

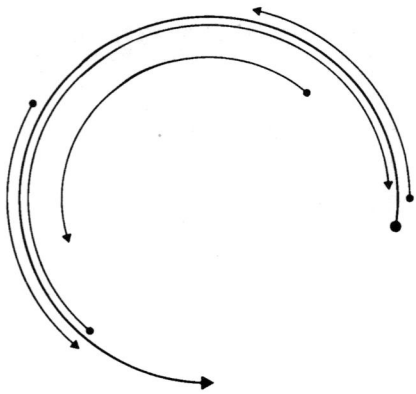

和另一个自己谈谈心

Talking with another self

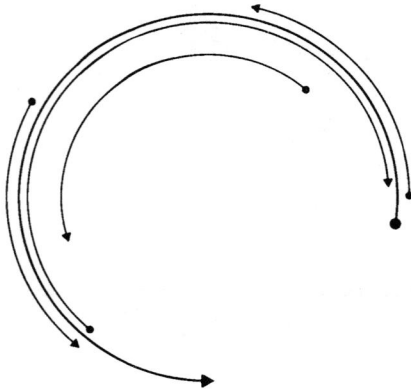

读一本书

Read a book

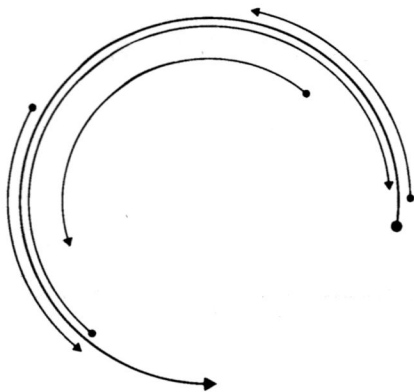

了不起的我

I am great

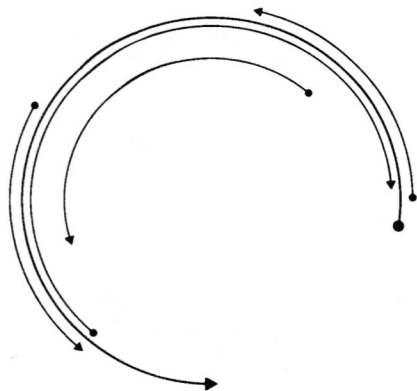

美好正在悄悄降临

Good is creeping in

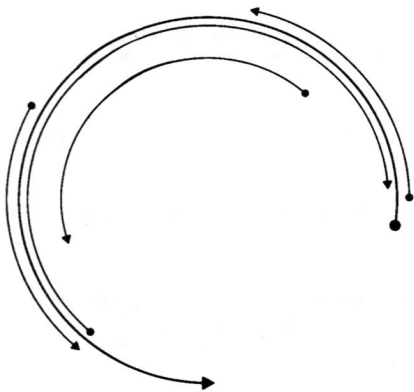

感谢此刻

Be grateful to this moment

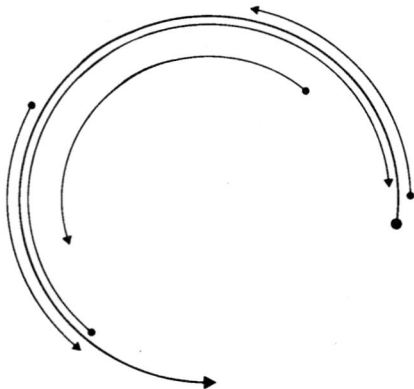

念念不忘，必有回响

What stays in your mind will someday spring up in your life

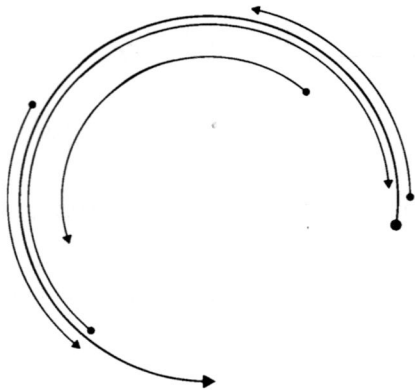

试着做出改变

Try to make a change

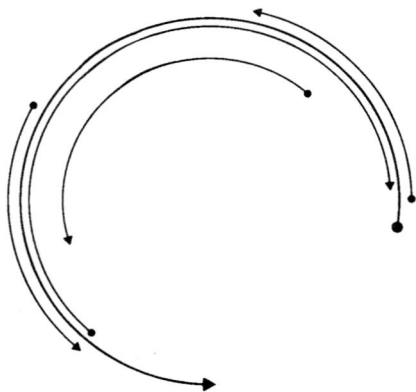

好好吃饭，按时睡觉

Eat well, sleep on time

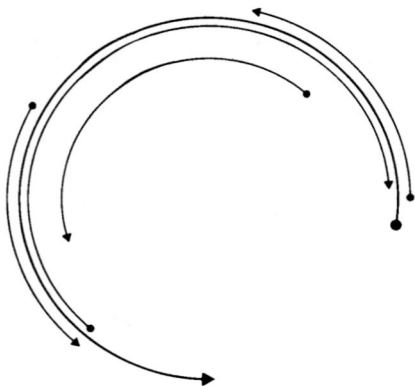

无谓的挣扎

Unnecessary struggle

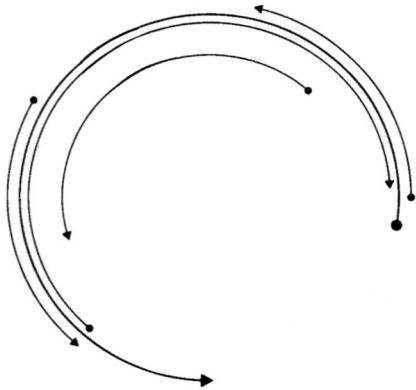

总会有人来爱你

There will always be someone to love you

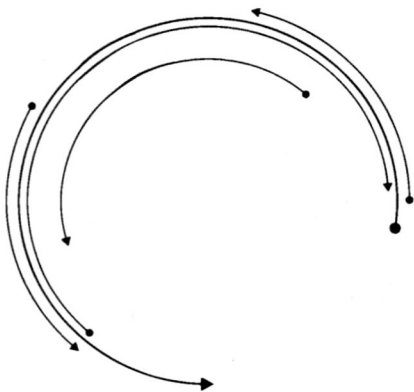

別再等了

Don't wait any longer

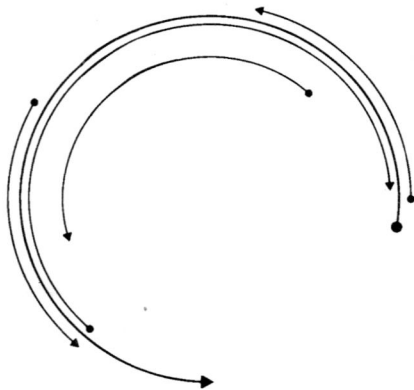

你可能会后悔

You might regret it

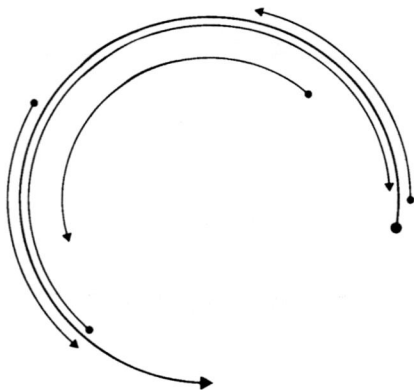

毫无疑问

No doubt

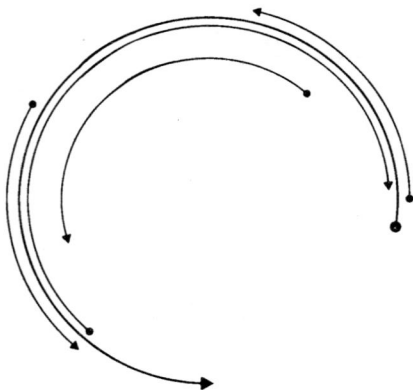

是的

Yes

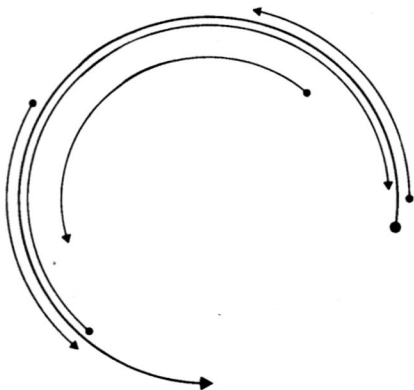

不

No

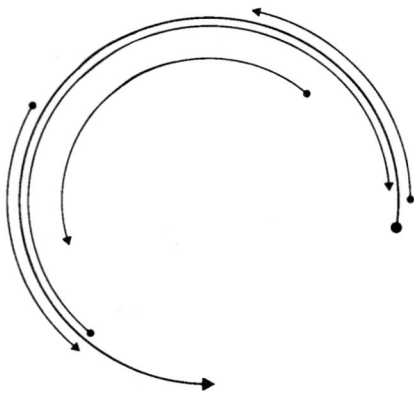

一切自有安排

Everything happens at its own pace

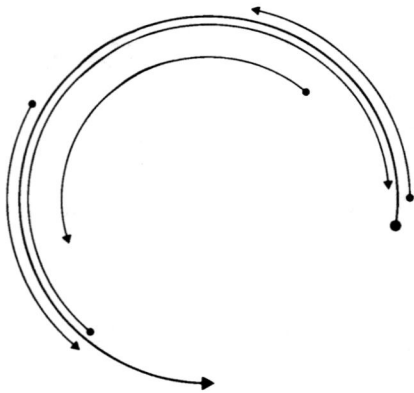

不要想得那么复杂

Don't over-complicated

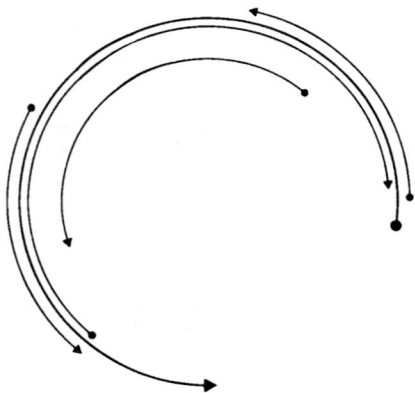

相信

Believe in it

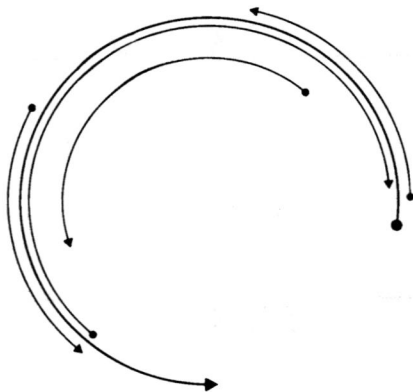

断舍离

Simplifying your life

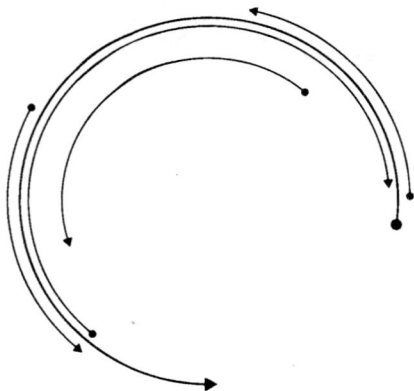

允许错误存在

Allow errors to exist

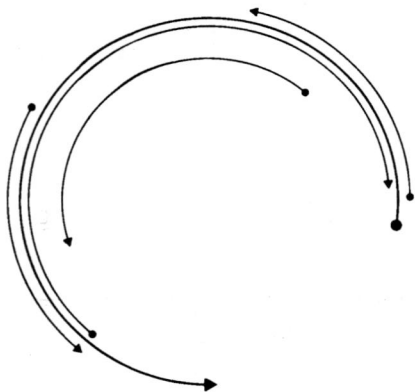

成长内在

Growing your inner being

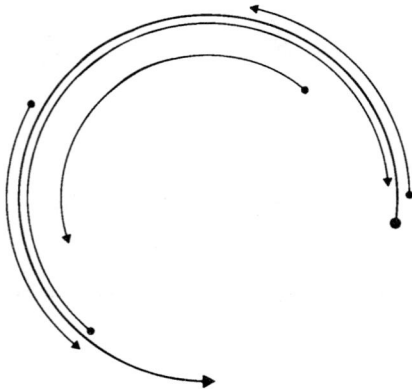

询问长者

Ask the elders

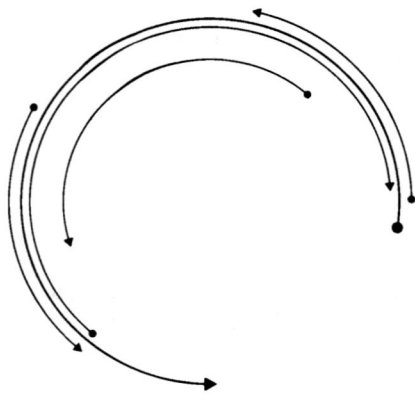

联系所有人

Connect with everyone

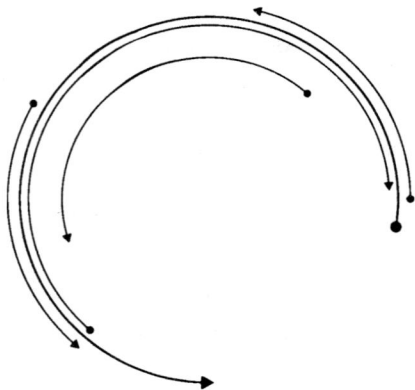

那些困难的都自有价值

Those difficulties have their own value

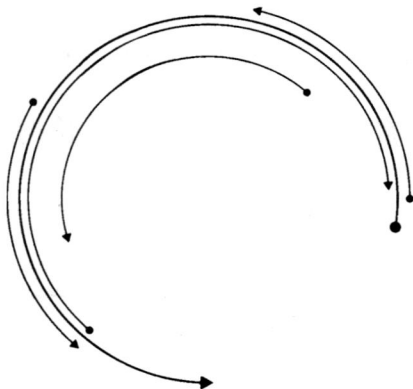

保持好奇心

Keep curious

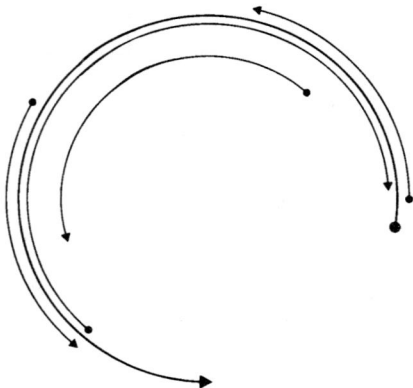

勇于探索

Dare to explore

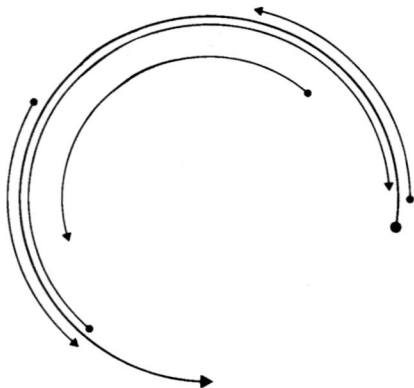

有可能会出现差错

There is possibility of error

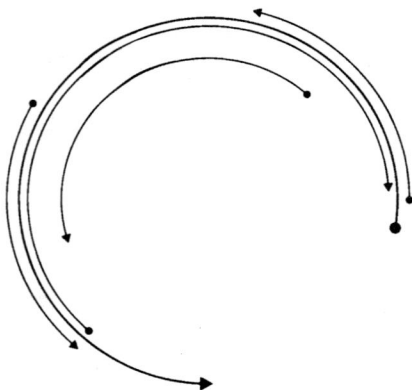

也许会遭到反对

You might be opposed

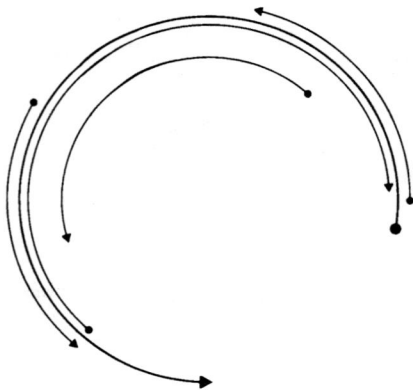

绝不要

Never

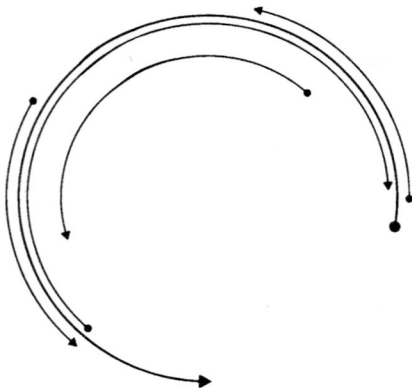

负起责任

Take responsibility

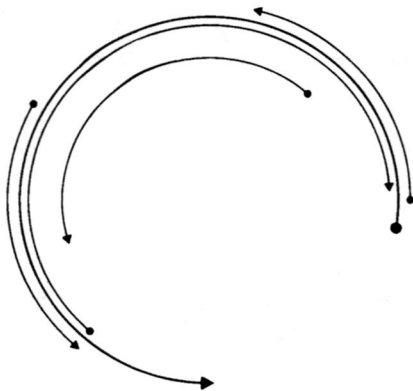

三思而后行

Look before you leap

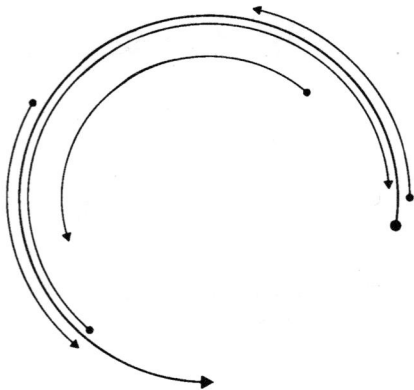

默数 1-10，重新提问

Count 1-10 silently and ask again

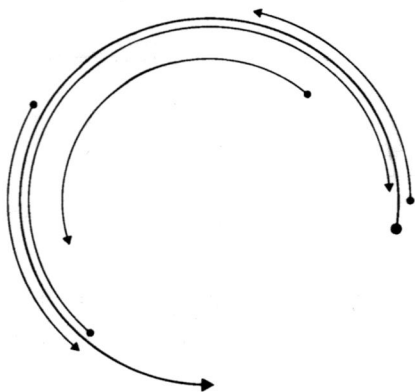

学会合作

Learning to cooperation

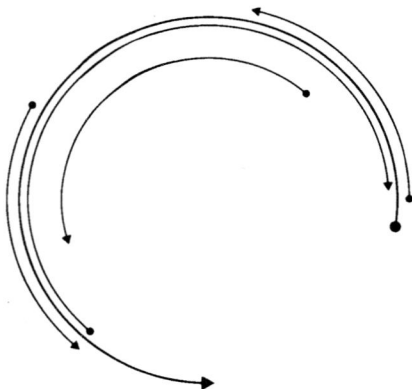

认真倾听

Listen carefully

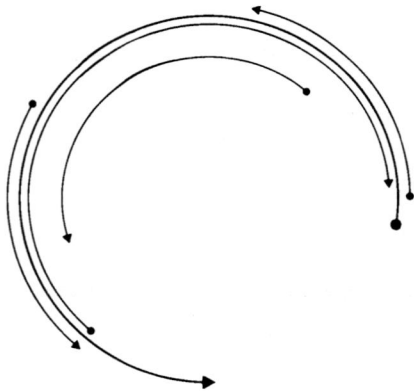

想想其他方法

Think of another way

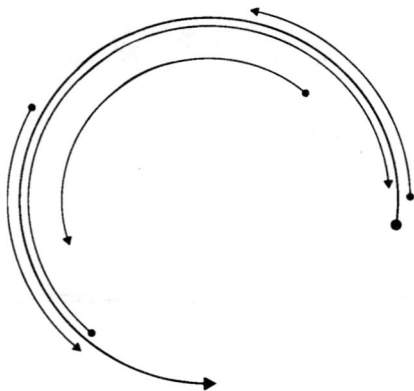

要么全力以赴做好，要么不做

Either go all out to do a good job or don't

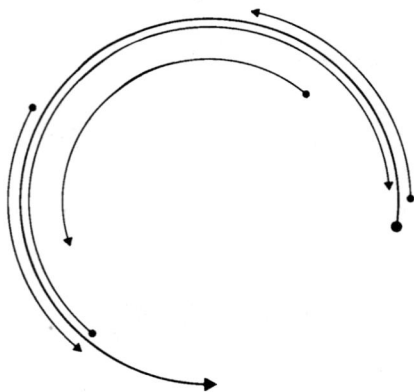

享受过程

Enjoy the process

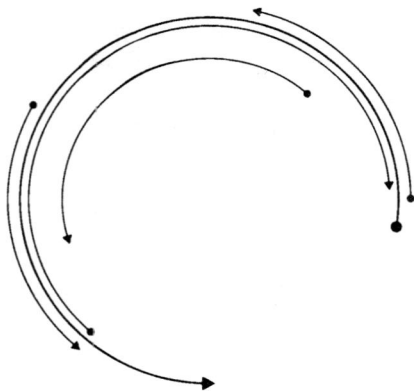

经历也是一种收获

Experiences are also rewards

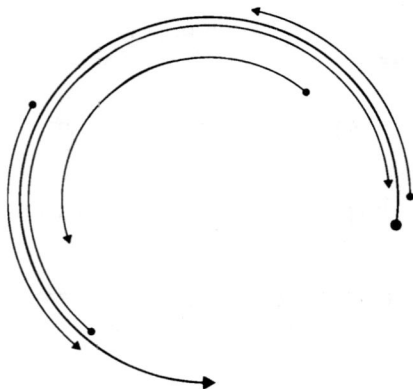

不要强迫

Don't force

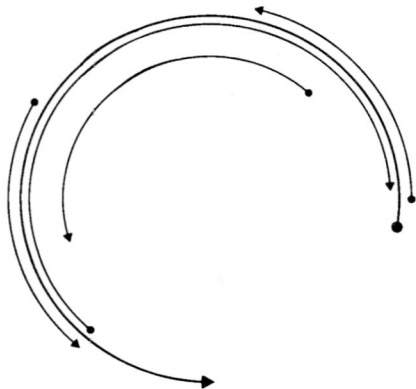

现在就行动

Take action right now

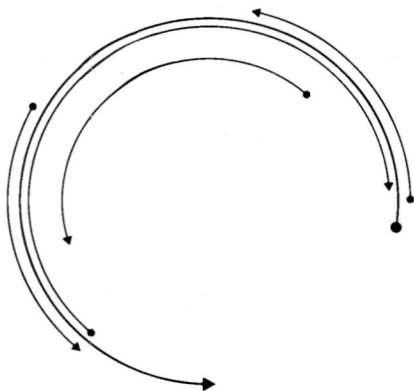

找一张纸，列出原因

Get a piece of paper and list the reasons

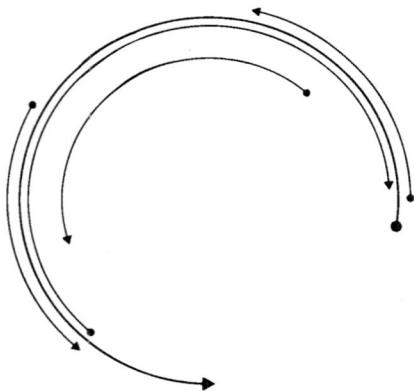

告诉他人这对你的意义

Tell others what this means to you

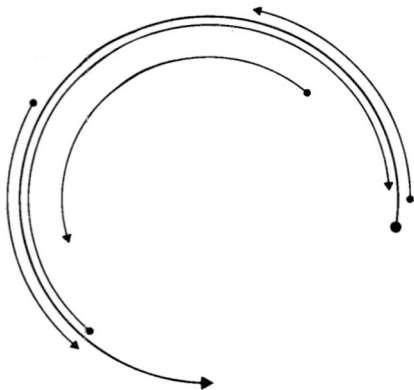

学会适应

Learn to adapt

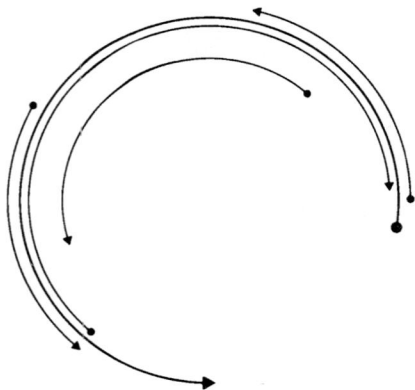

终将做出妥协

Compromise will be made eventually

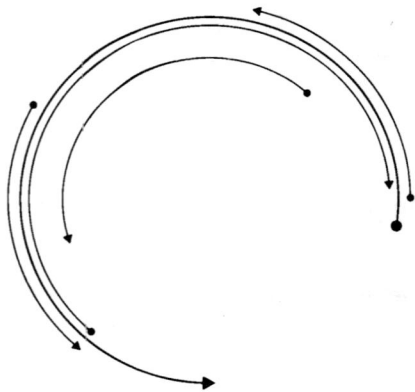

毋庸置疑

Indubitable

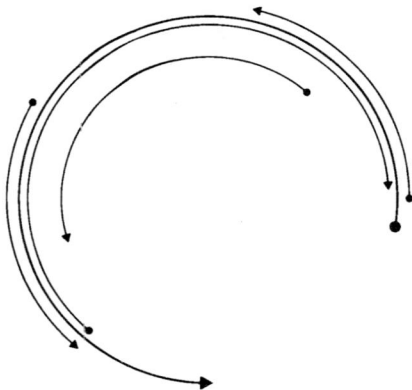

当作一次机会

Consider it as an opportunity

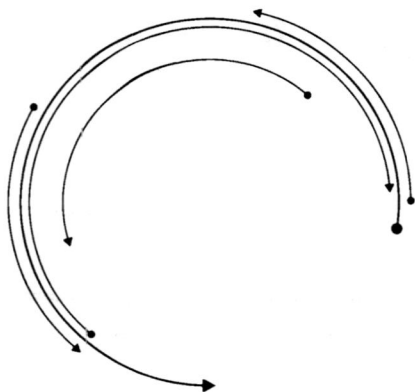

这是不明智的

It is unwise

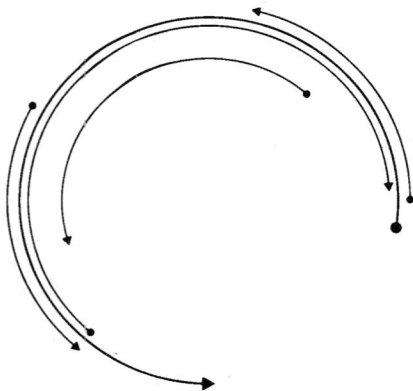

关注细节

Pay attention to detail

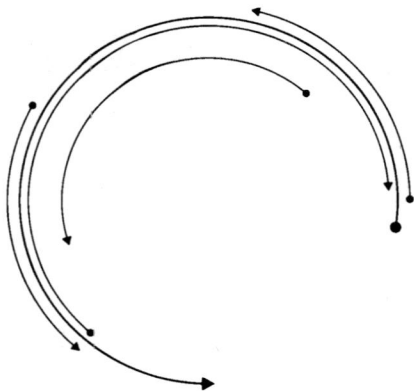

提高注意力

Improving concentration

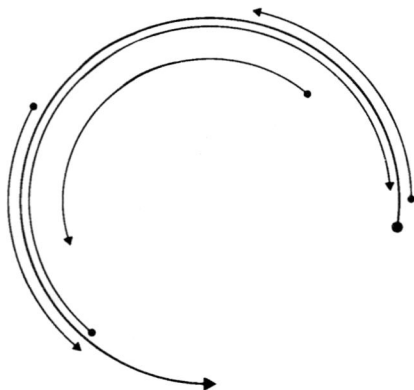

这是在浪费时间

It's a waste of time

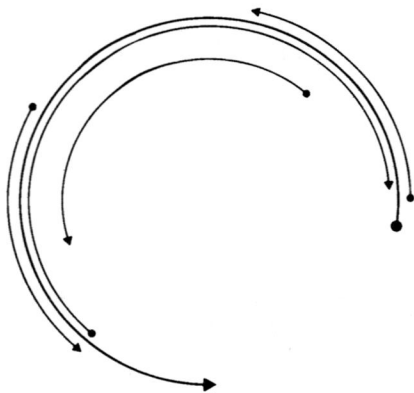

也许

Maybe

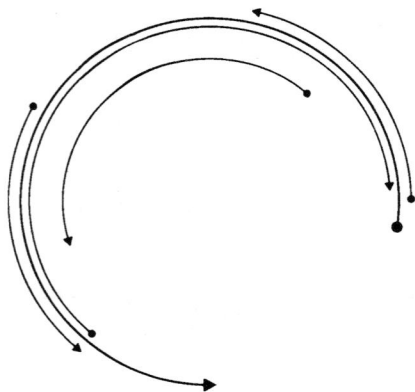

你会失望的

You will be disappointed

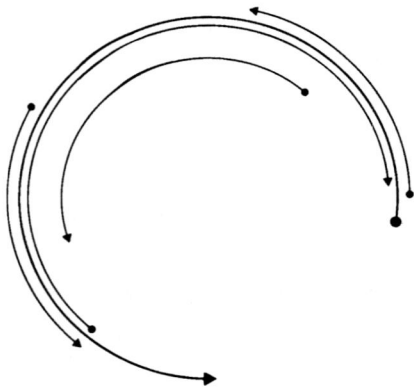

那会是一件有趣的事儿

That would be interesting

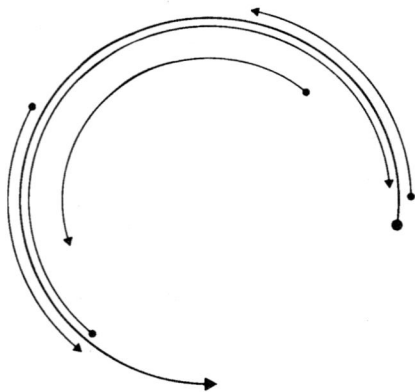

保持开放的心态

Keep an open mind

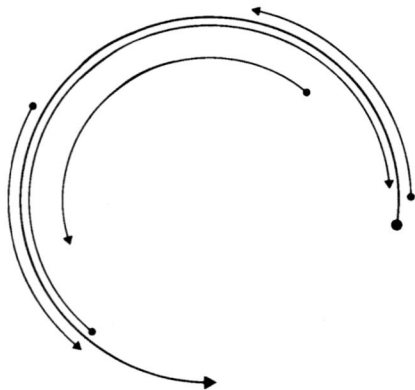

别用习惯的方式去解决

Don't solve it in a habitual way

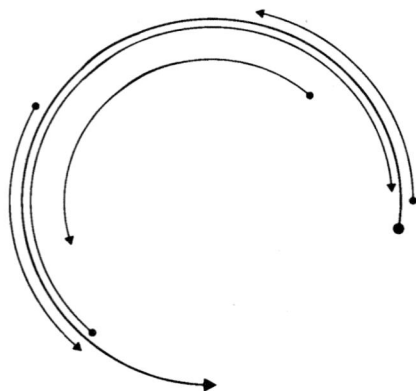

向前看

Look forward

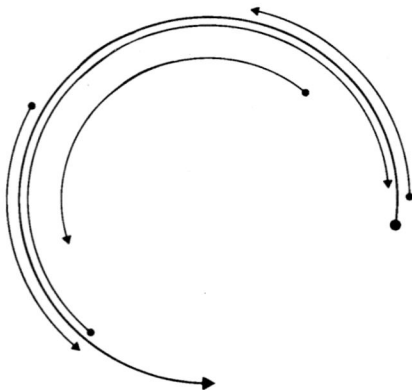

尽早开始

Start as early as possible

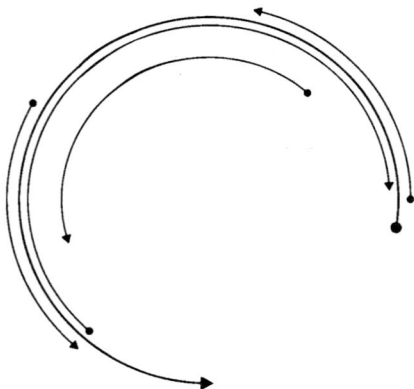

当下就是最好的时光

The present time is the best time

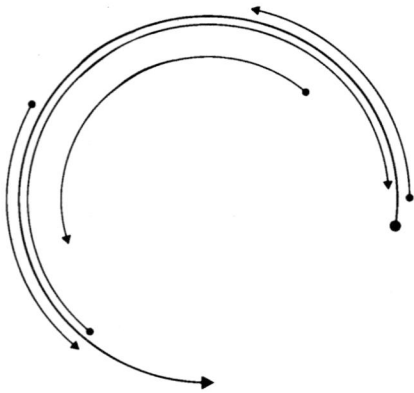

别太在意

Don't take it too seriously

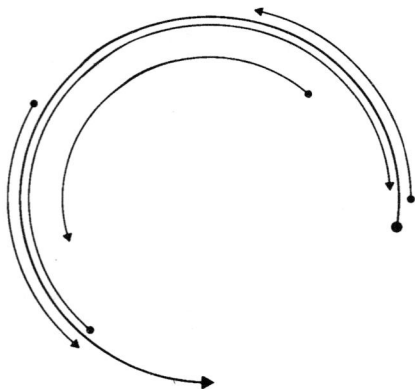

不要奢望

Don't have extravagant hopes

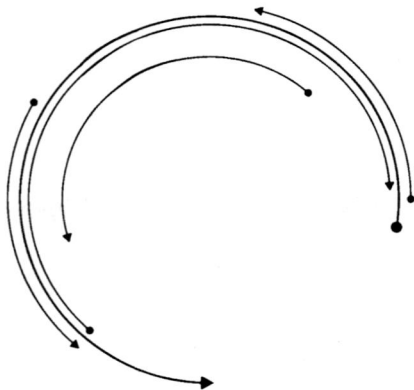

不适合在这个时候

It's not suitable at this time

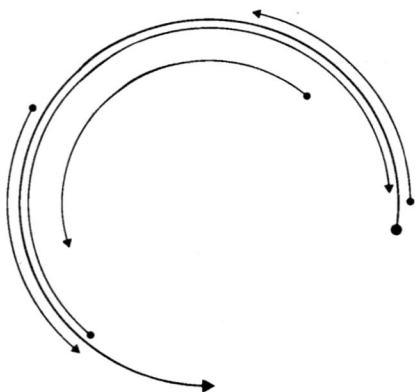

看看下一页

Take a look at the next page

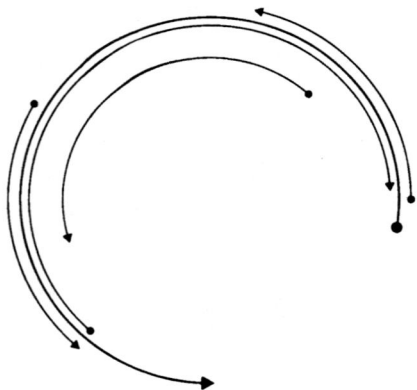

终将得偿所愿

You will finally get what you want

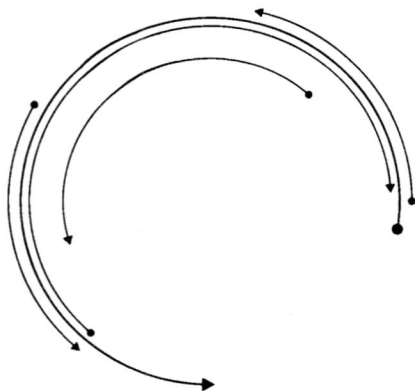

你的行动会带来转机

Your actions will take a turn for the better

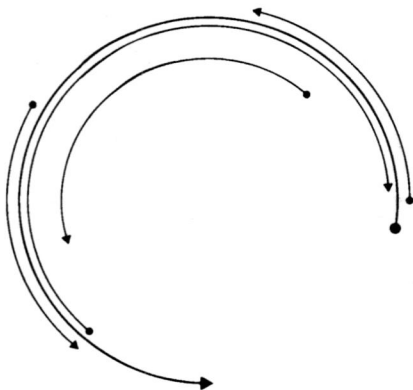

付出努力是值得的

The effort is worth it

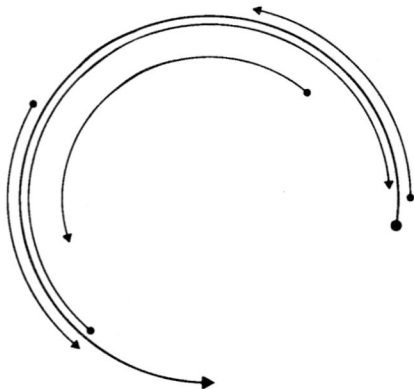

别嫌麻烦

Don't mind the trouble

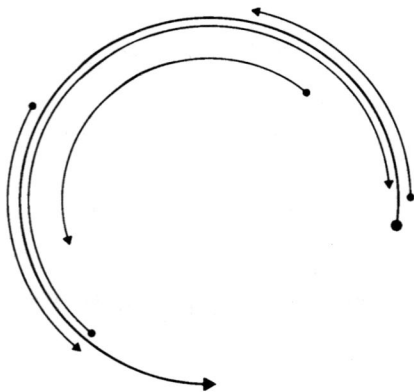

晚一些再做决定

Make a decision later

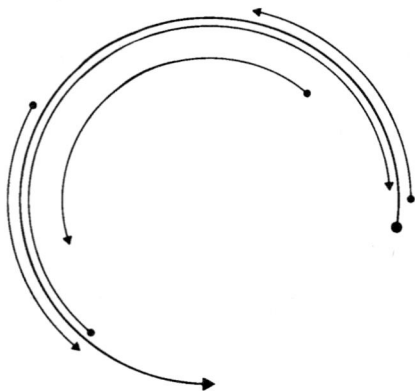

冥想

Meditate

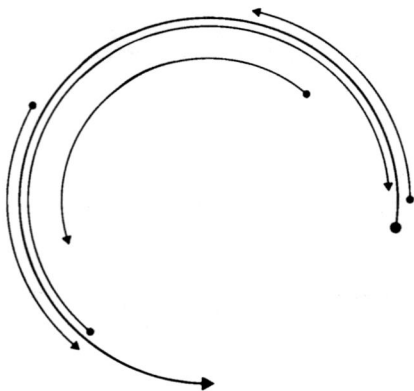

这是一个改变自己的好机会

It's a good opportunity to change yourself

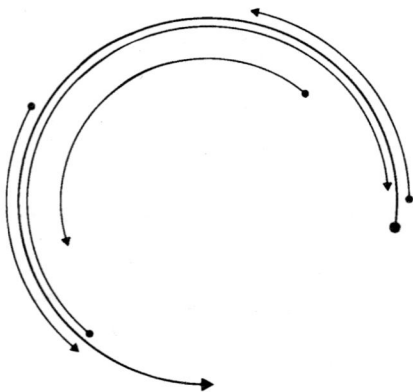

无解

Unsolvable

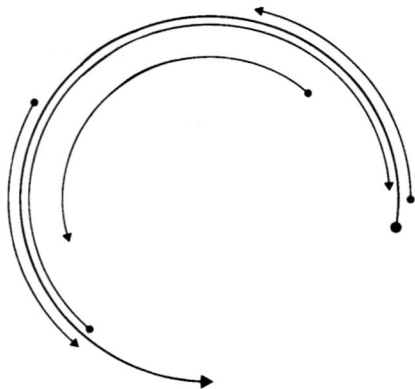

微妙的平衡

A delicate balance

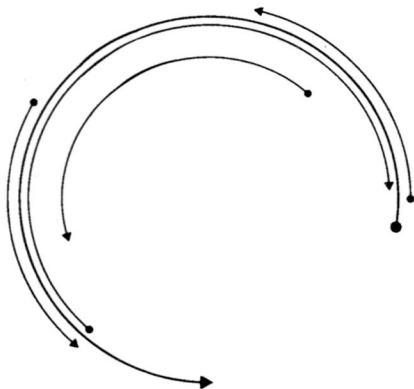

需要一个新计划

Need a new plan

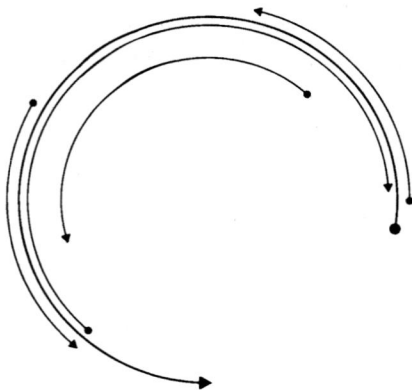

任何选择都不能保证全是好结果

No choice can guarantee good results

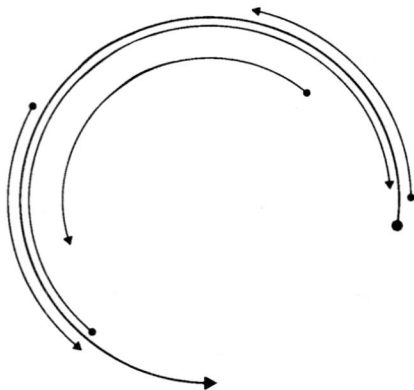

这并不重要

It doesn't matter

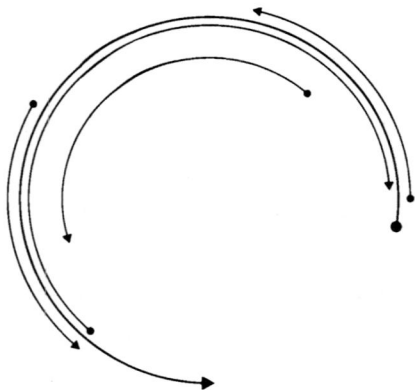

别 用 第 一 个 想 到 的 解 决 方 案

Don't use the solution that comes to mind at the first time

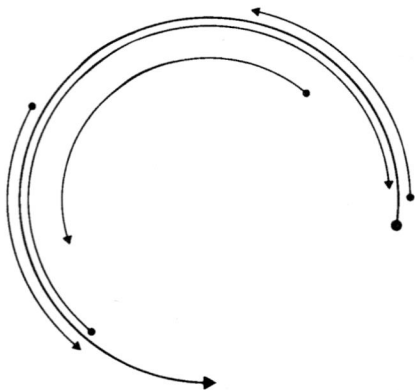

要有耐心

Be patient

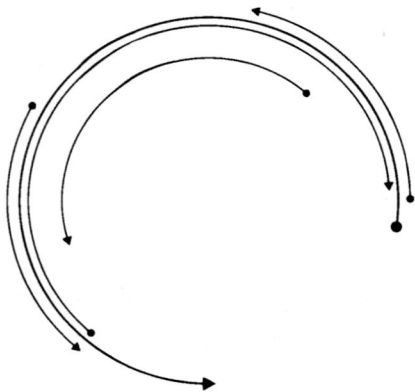

有可能会引发状况

It may causes an unexpected situation

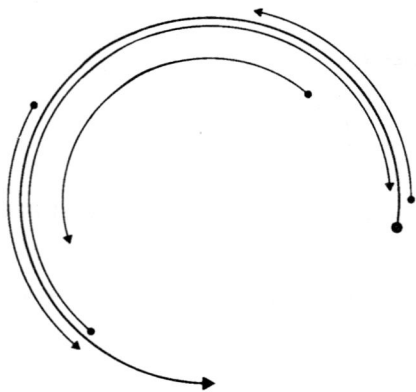

坚持你的决定

Stick to your decision

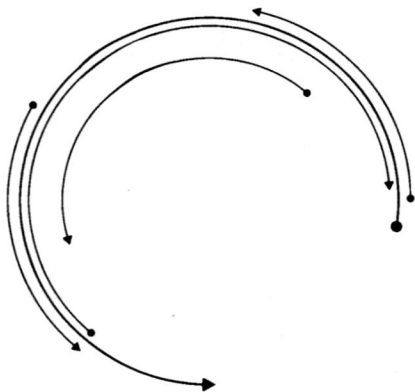

居心叵测

With an ulterior motive

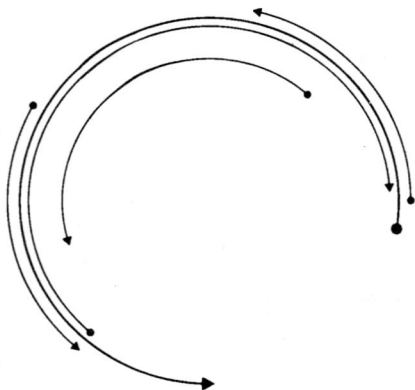

命运

Destiny

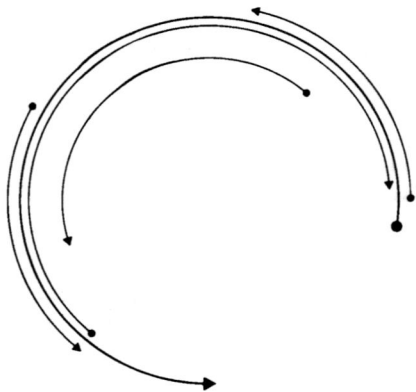

握住他人伸出的手

Hold the outstretched hand of others

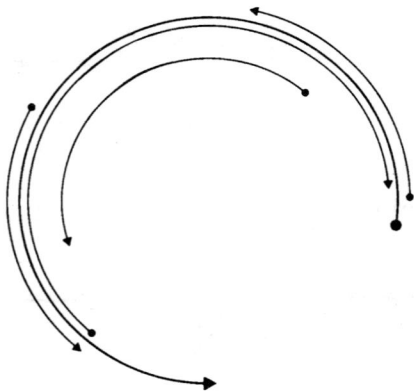

走出剧情

Step out of the plot

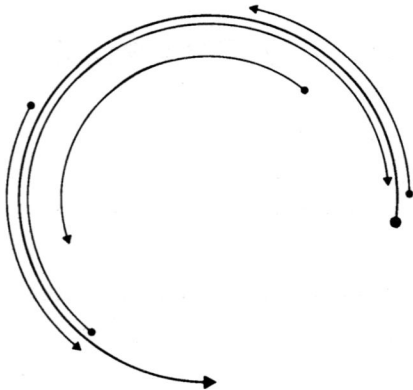

最重要的事，只有一件

The most important thing is only one

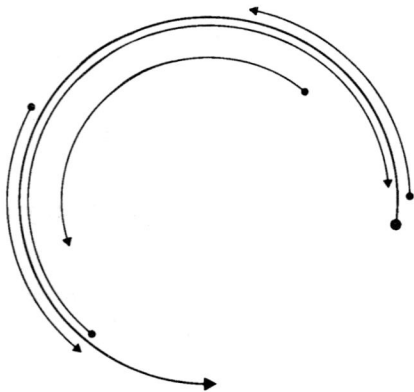

会得到支持

Will be supported

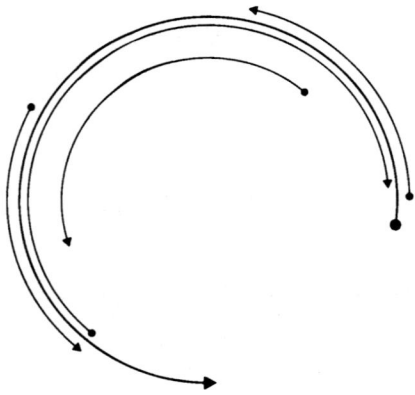

做好自己该做的

Do what you should do

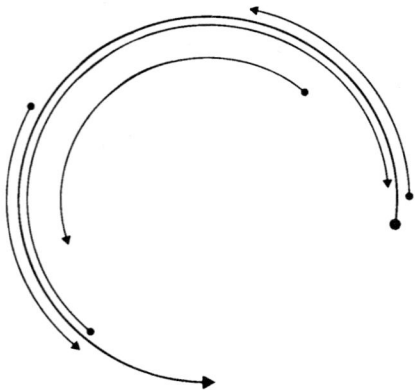

不要再继续问了

Don't keep asking

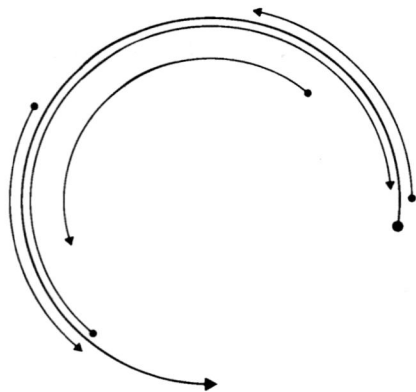

记录它

Record it

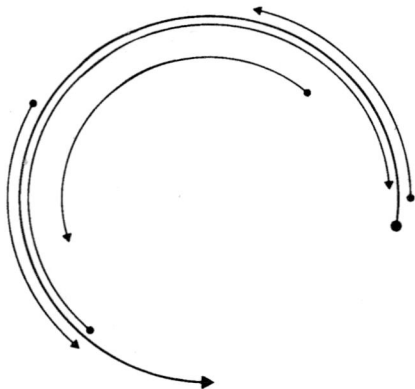

也许该找个人聊聊

Maybe it's time to talk to someone

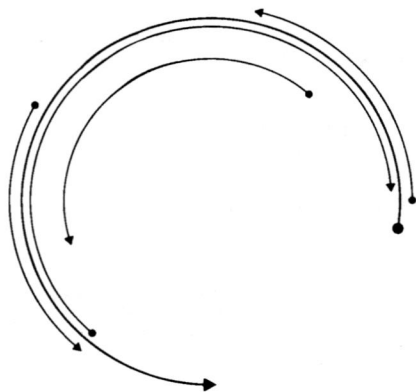

默默无闻

Unknown to public

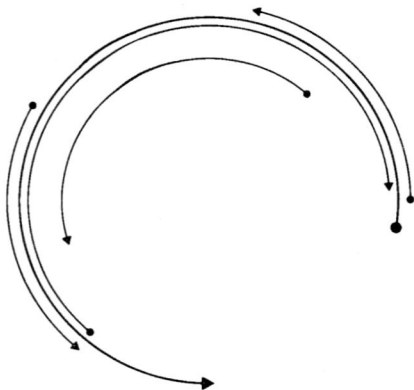

学会设置优先级

Learn to set priorities

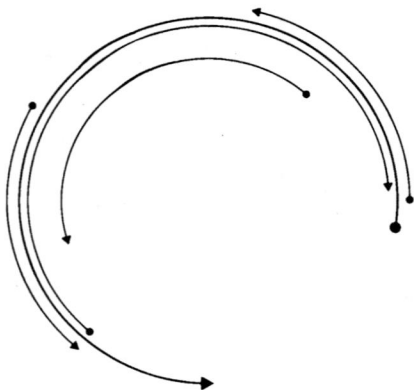

先完成其他事

Finish other things first

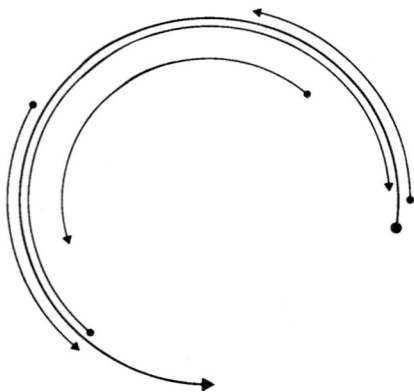

可能不得不放弃其他东西

Perhaps have to give up something else

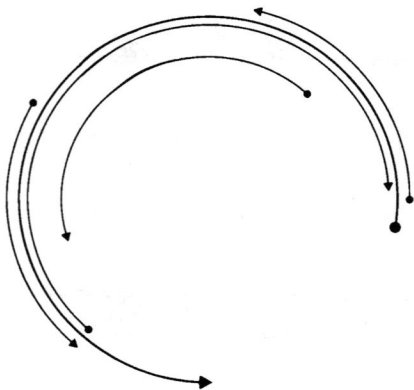

方法就在眼前

The method is right in front of your eyes

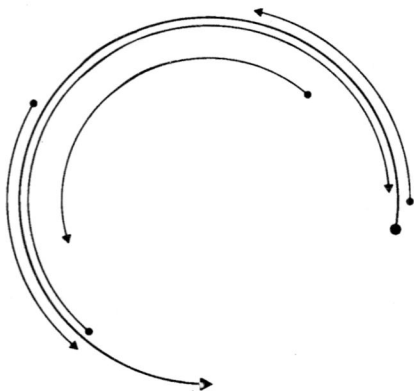

这取决于你的态度

It depends on your attitude

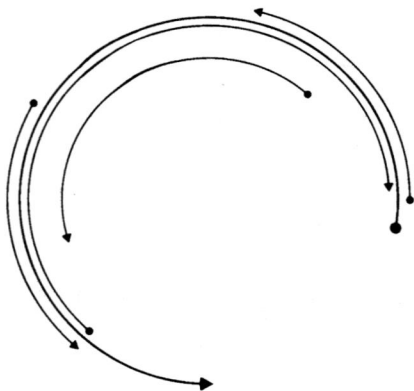

当然

Of course

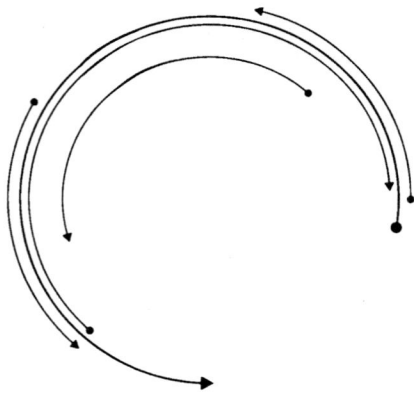

似乎就是这样

It seems like this

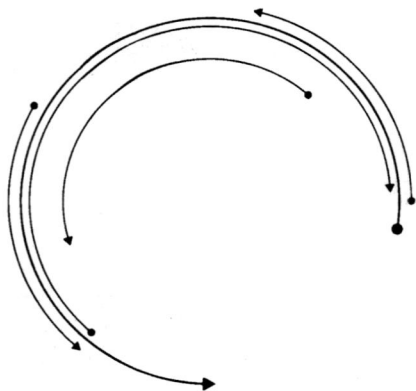

你并不孤独

You are not alone

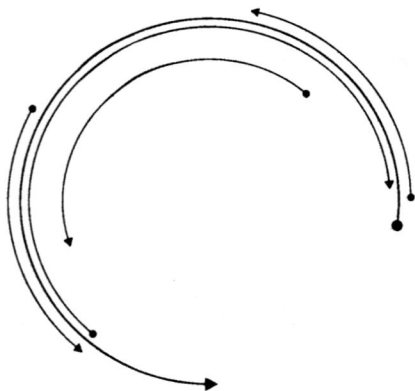

有些事情显而易见

Some things are obvious

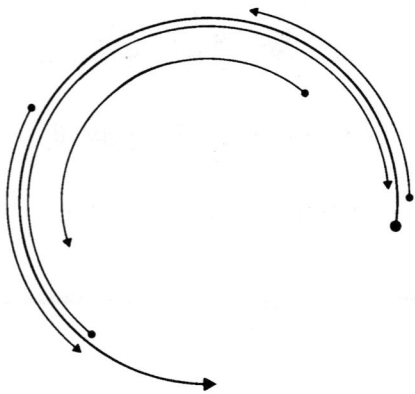

接受援手会发展得更好

Accept assistance will leads to better development

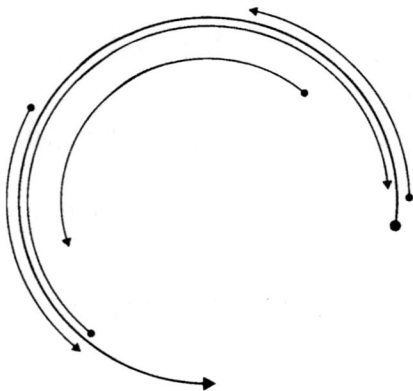

它会帮你熬过去

It will carry you through

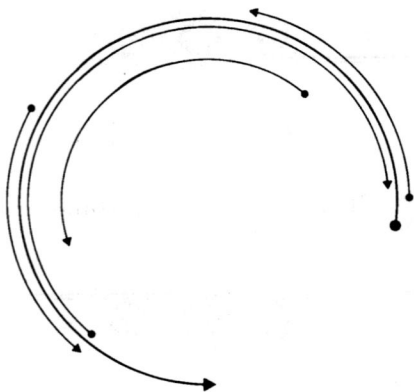

真的无法妥协

Really hard to compromise

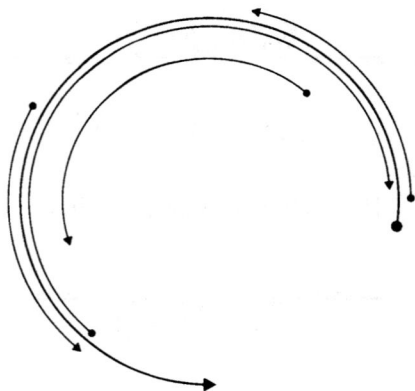

保持沉默

Keep silent

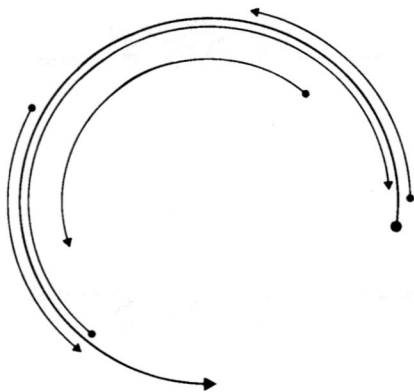

静心

Calm the mind

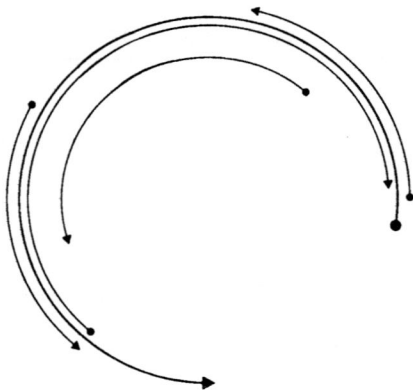

大方说出来

Say it out liberally

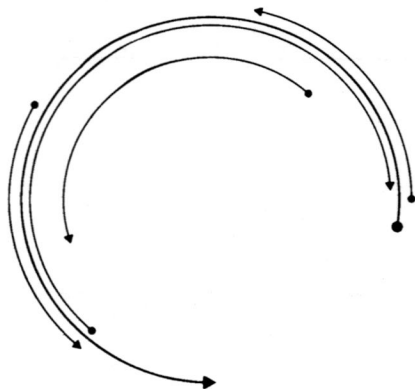

无关紧要

It's beside the point

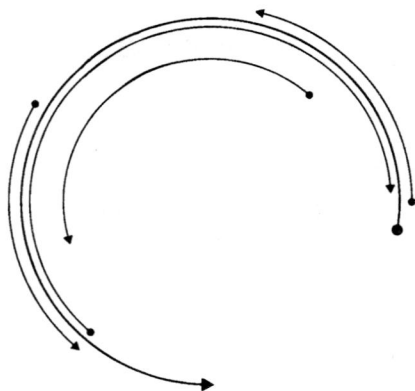

一年以后再看

Check back in a year

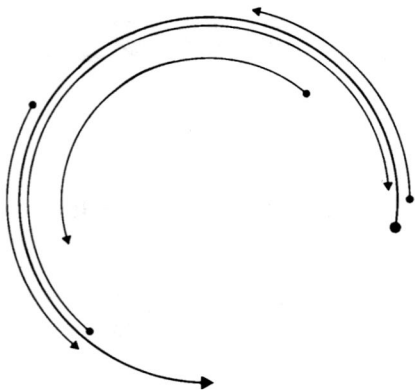

这会非常美好

This is going to be wonderful

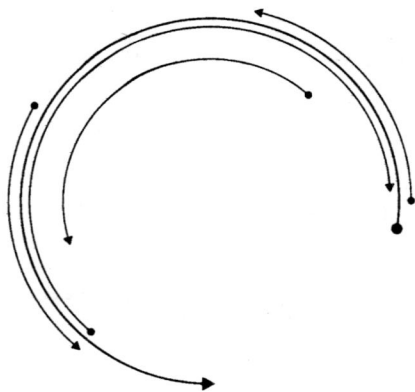

无法保证

This is unwarranted

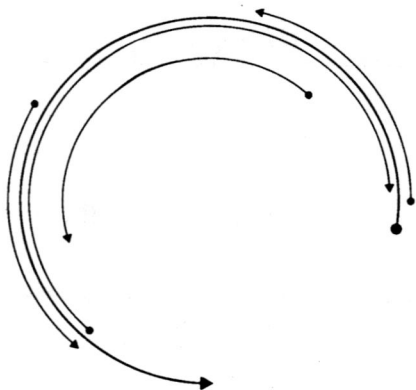

那将超出你的控制范围

That would be beyond your control

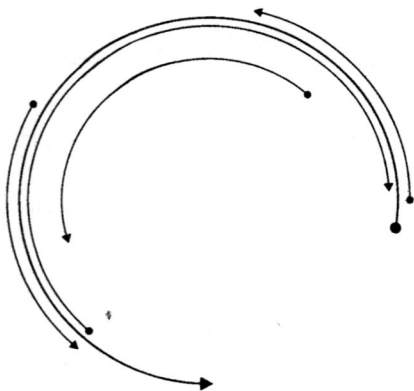

坚持下去

Keep going

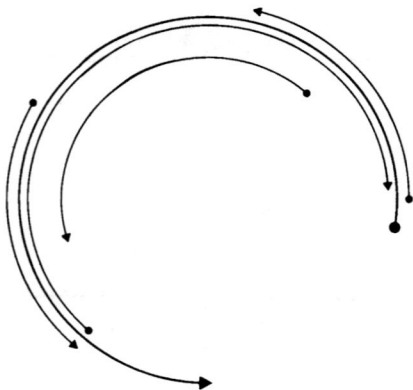

温柔地反对

Tenderly opposed

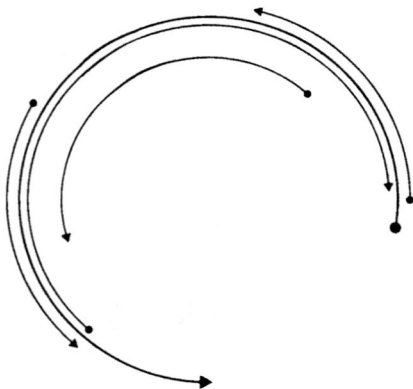

保持灵活度

Stay flexibility

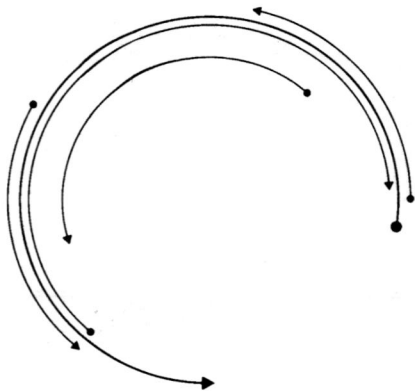

延迟满足

Delay gratification

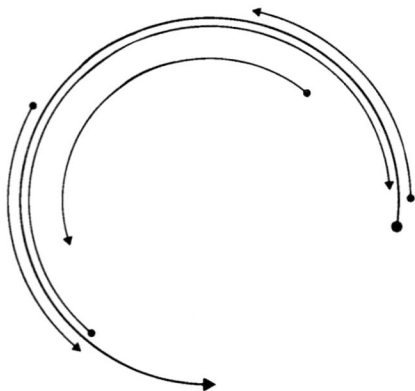

如果一切如你所说

If everything is as you said

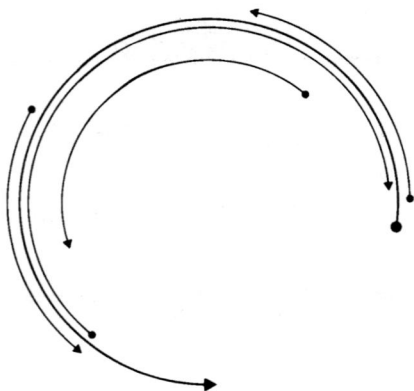

不要冲动

Don't be impulsive

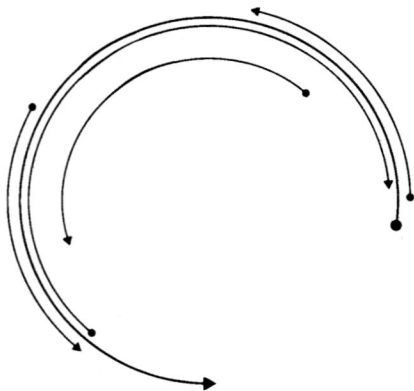

换个角度想想

Think about it from a different perspective

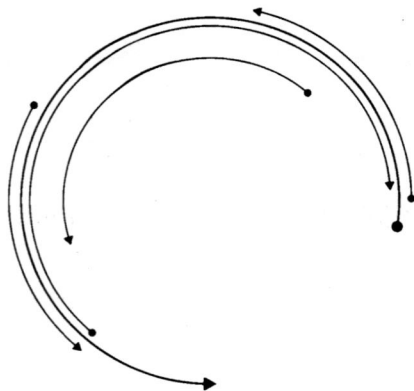

征求身边人的意见

Ask for the advice of those around you

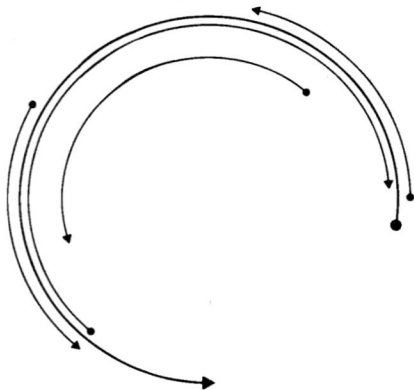

别怀疑

Don't doubt it

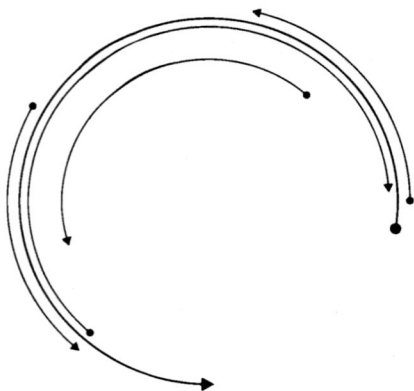

当你更成熟一些

When you are a little more mature

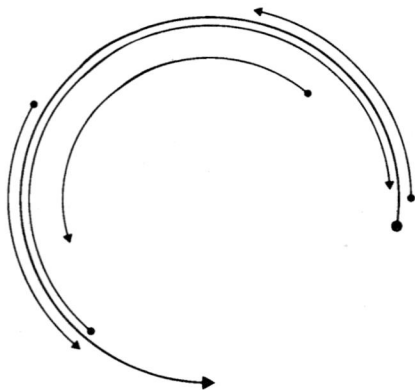

这是你可以决定的

It's up to you

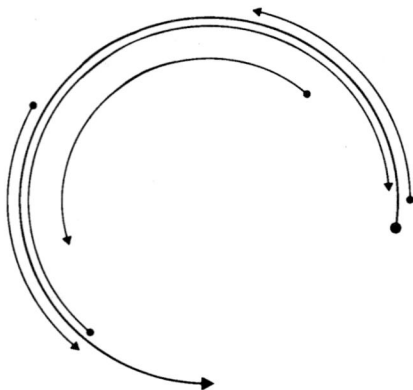

偶尔懈怠

You can occasionally slack off

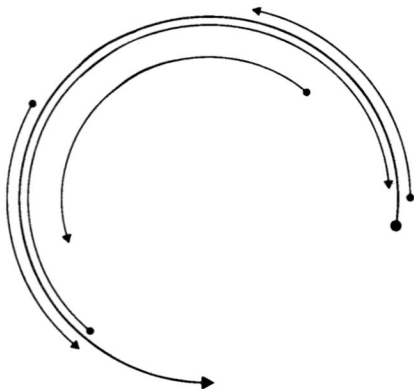

抱住棒棒的自己

Embrace the best of yourself

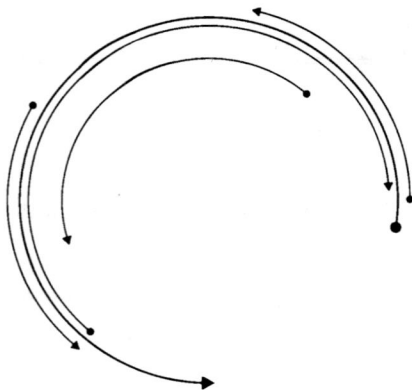

有得必有失

When there is gain, there is loss

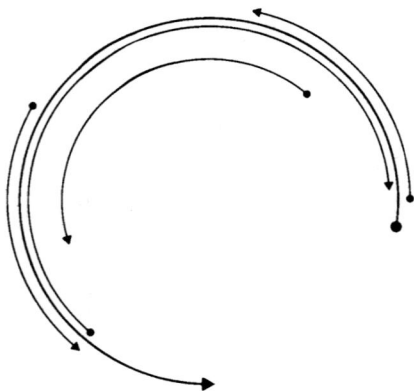

善于运用想象力

Make good use of imagination

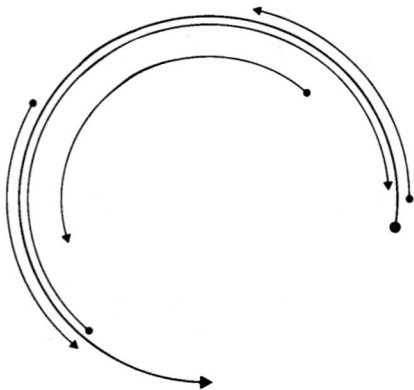

这不可预测

This is unpredictable

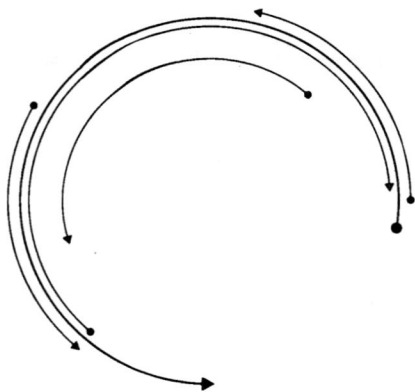

别太焦虑

Don't be too anxious

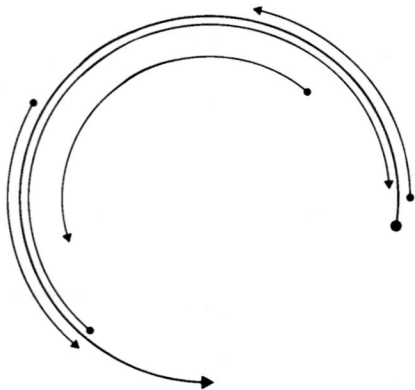

过去没那么重要

The past is not so important

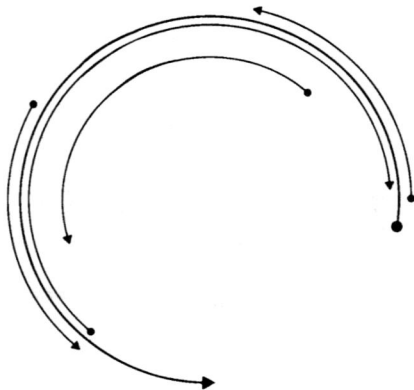

学会自得其乐

Learn to enjoy yourself

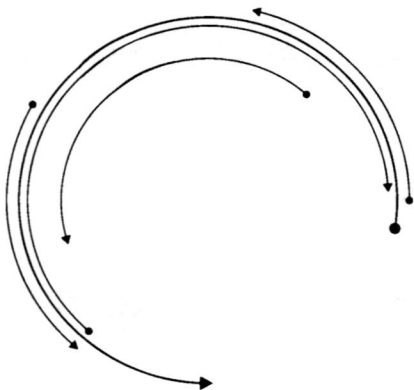

谨言

Speak cautiously

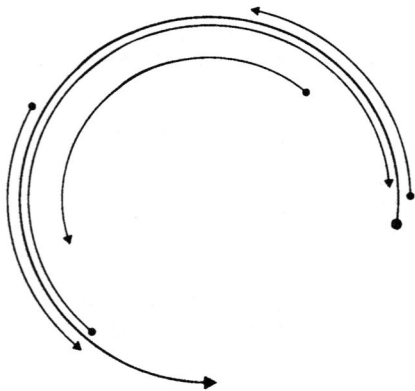

这是可以做到的

You can make it

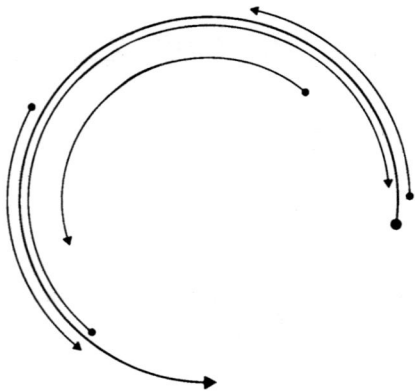

当断不断，反受其乱

Indecisiveness leads to disaster

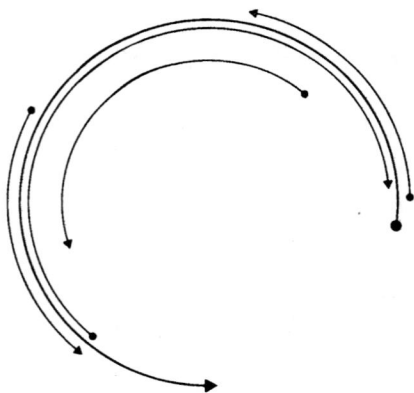

不相上下

Neck and neck

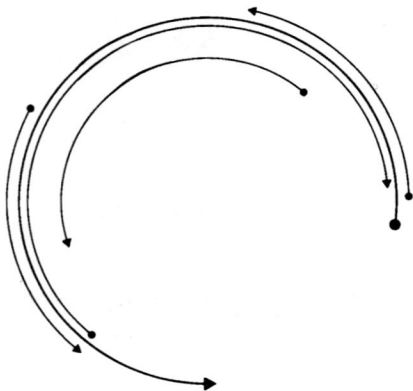

重新审视什么是重要的

Re-examine what matters

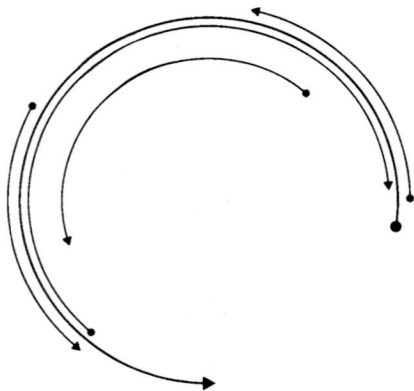

塞翁失马，焉知非福

It's a blessing in disguise

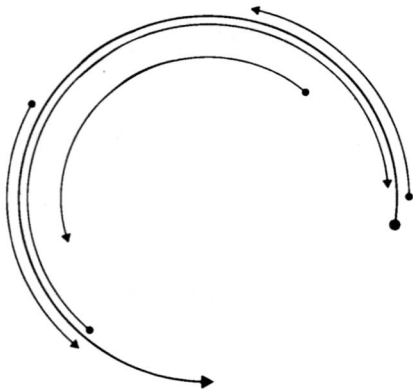

刻意练习

Deliberate practice

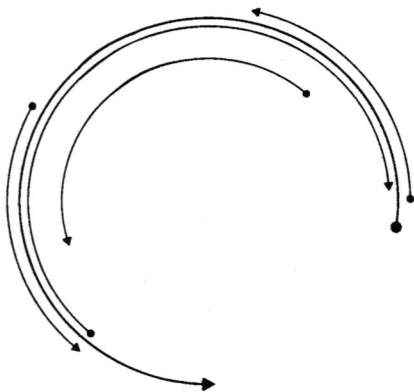

结局值得期待

The ending is worth expecting

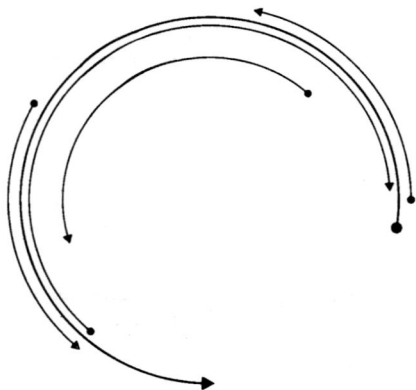

微笑面对

Face it with a smile

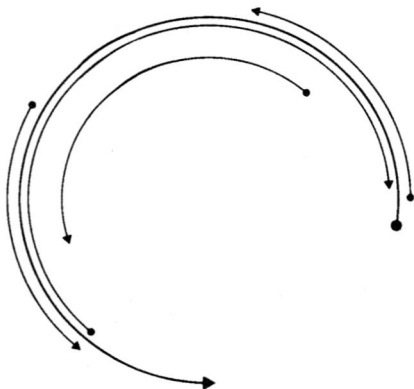

去问问父母

Ask your parents

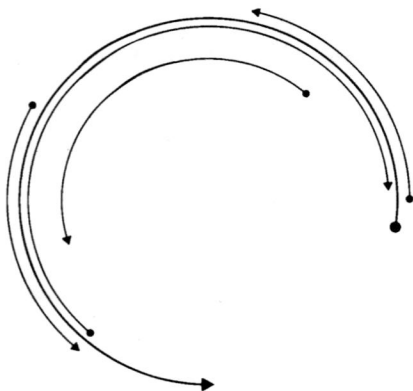

打不垮你的才会成就你

What doesn't beat you makes who you are

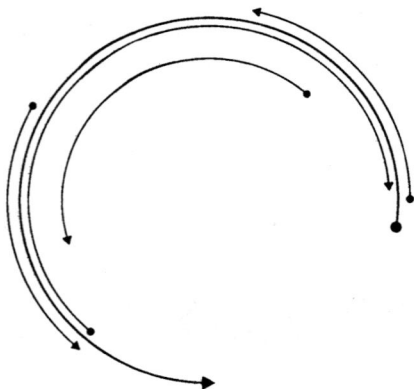

烦恼可能是自找的

You may suffer a self-inflicted trouble

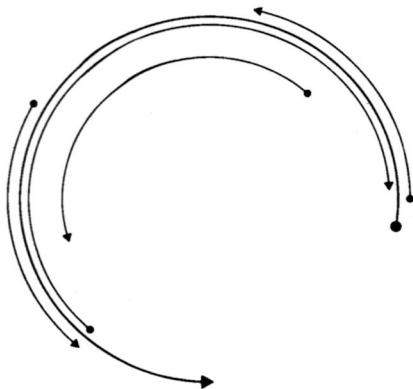

不要计较

Let it go

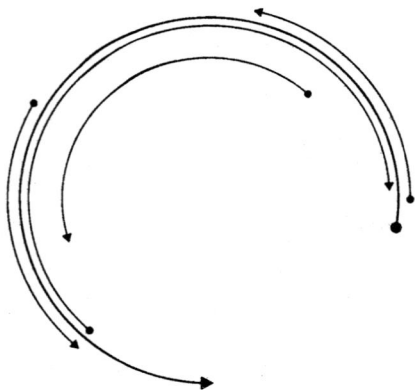

珍惜当下

You only live once

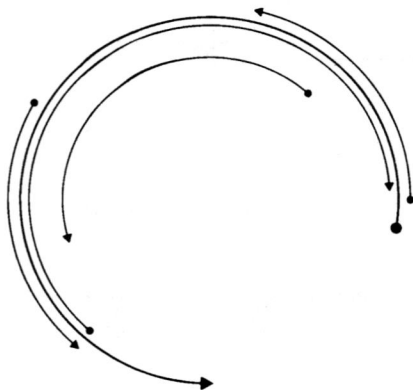

把目标写下来

Write down your goals

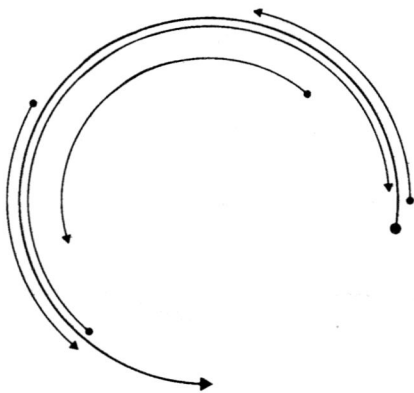

懂得转身而不是后退

Know how to turn around instead of drawing back

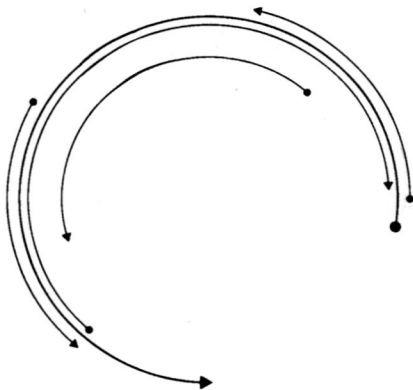

邯郸学步不可取

Don't imitate others and thus lose your own individuality

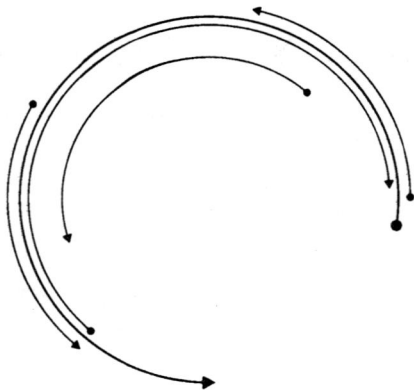

会忘掉的

It'll be forgotten

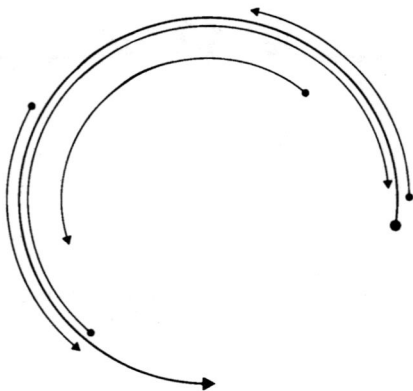

犯错未必带来恶果，但胡闹会

Making mistakes may not bring bad consequences,
but mess around can

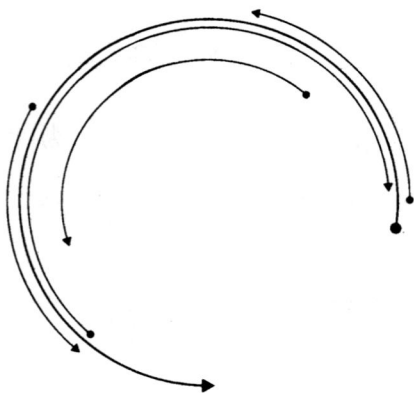

情绪也是一种力量

Emotion is also power

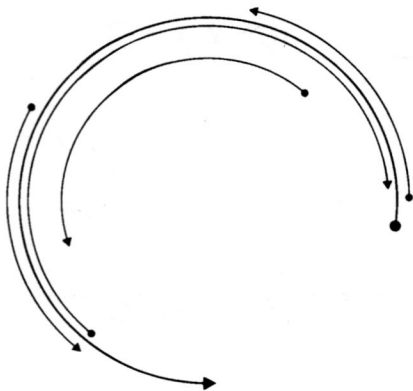

破圈

Think outside the box

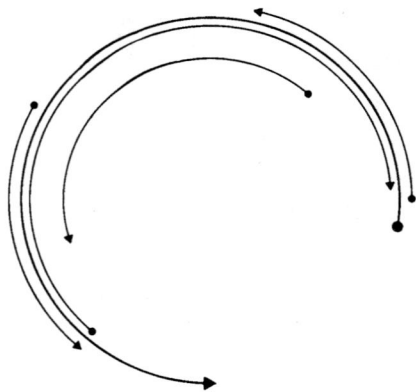

与自己和解

Reconcile with yourself

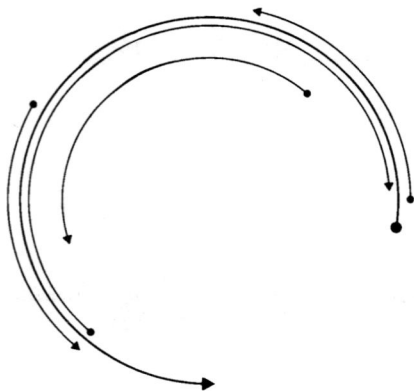

不存在十全十美

There is no perfection

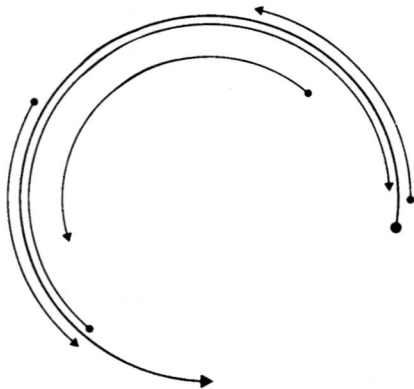

面对它

Face it

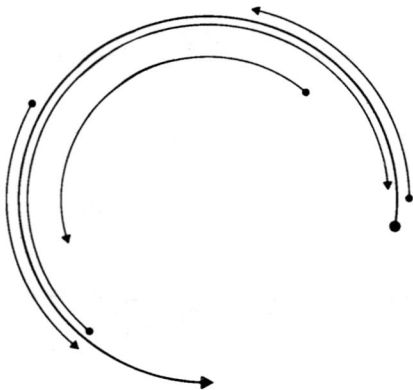

不要伤人伤己

Don't hurt others or yourself

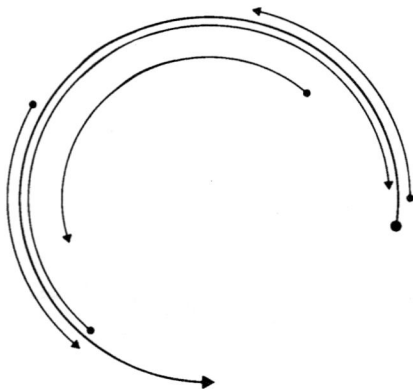

不要这样

Don't do this

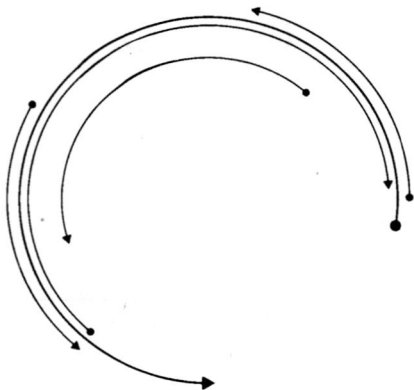

这也是一种考验

This is also a test

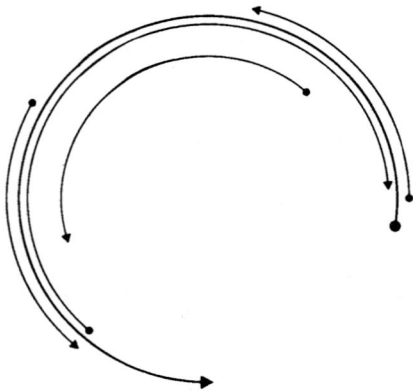

不要吝啬感激

Don't be stingy with gratitude

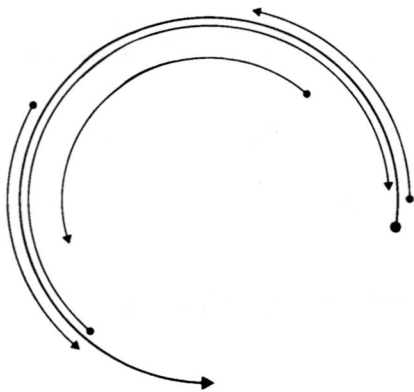

幸福是可以通过学习获得的

Happiness can be acquired through learning

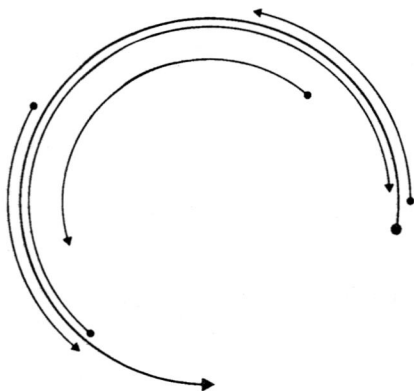

第二次受伤的机会

A second chance to get hurt

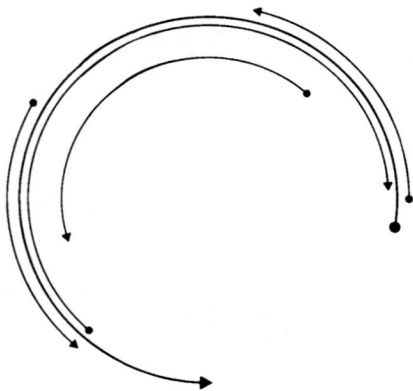

追随内心的声音

Follow your heart

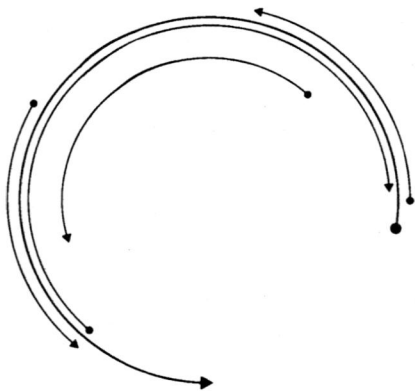

警惕习惯性思维

Be alert to habitual thinking

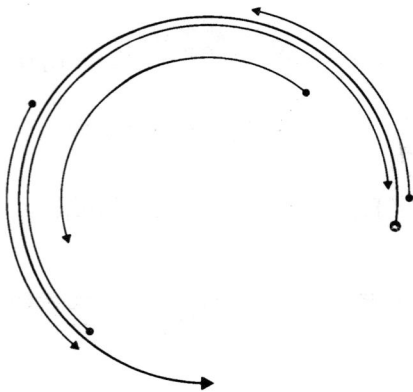

348

不耽于欲望

Don't pursue lust too much

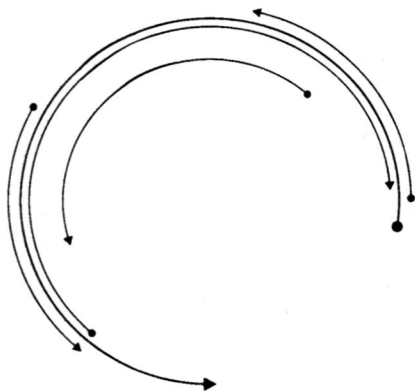

350

斯人如彩虹，有些事也是如此

Some people are iridescent,some things are the same

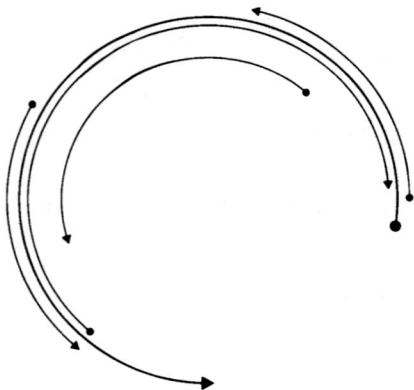

你追求的可能在变化

What you pursue may be changing

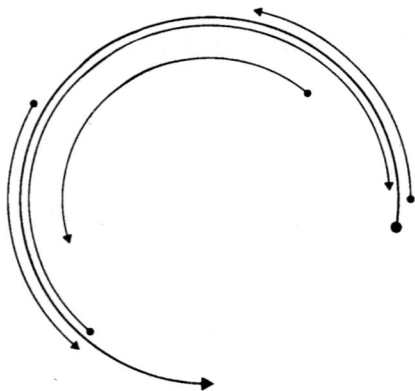

降低期待

Lower your expectations

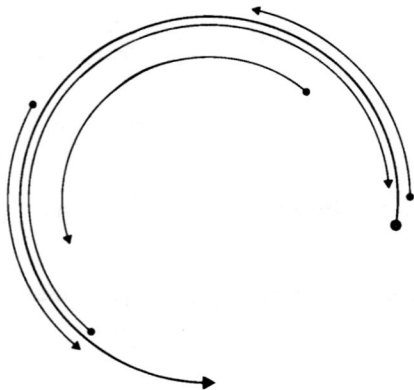

想要的未必都得到

You may not be able to get what you want

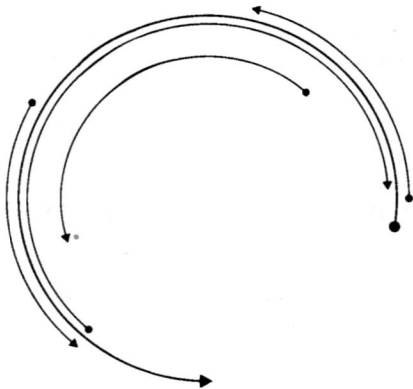

保持动力

Stay motivated

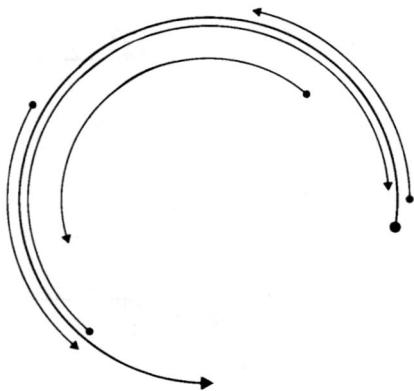

没有绝不会后悔的选择，只有选择不去后悔

There is no choice that will never regret,
only choose not to regret

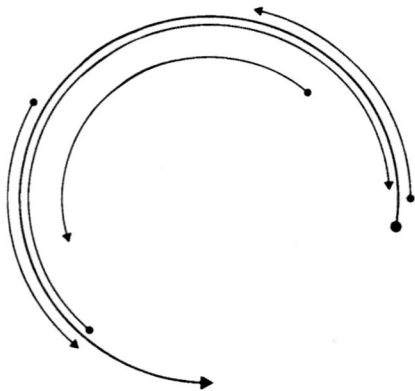

尽快开始

Start as soon as possible

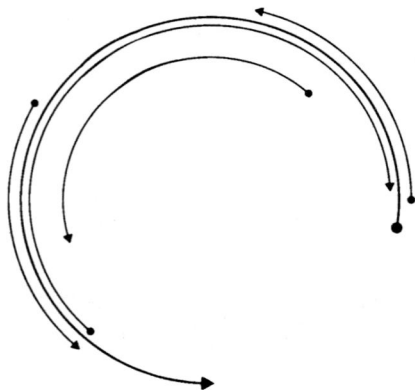

接受不确定性

Accept uncertainty

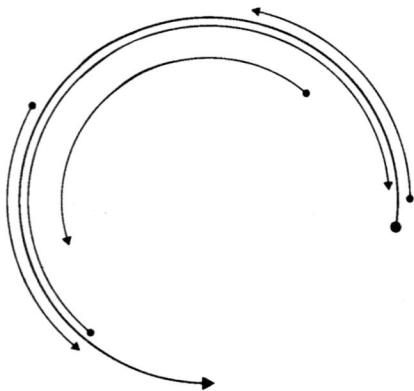

不要总是期待奇迹

Don't always expect a miracle

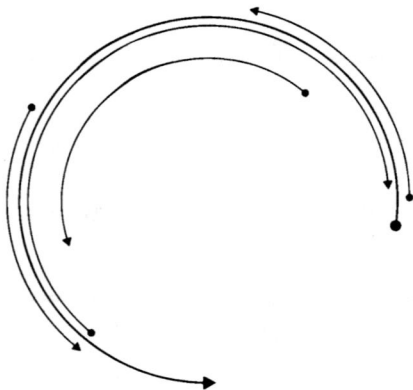

你想知道的终会了解

What you want to know will eventually be understood

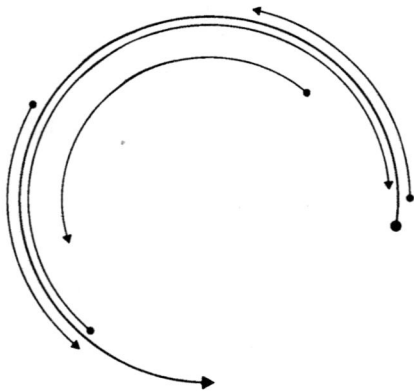

存在竞争

There is competition

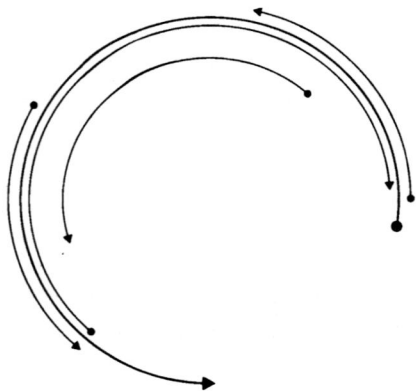

呵护

Care

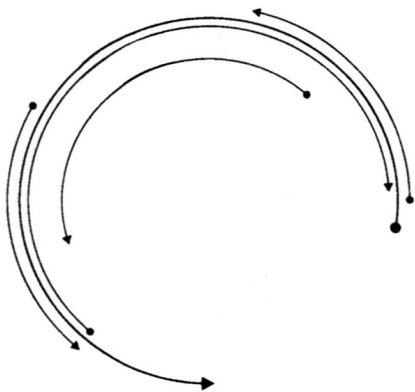

你可能已经拥有真正重要的东西

You may already have something really important

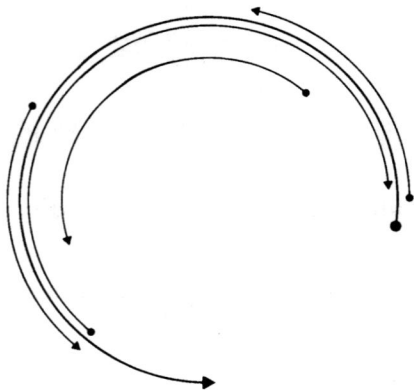

抱怨在所难免，但它毫无意义

Complaining is inevitable, but it makes no sense

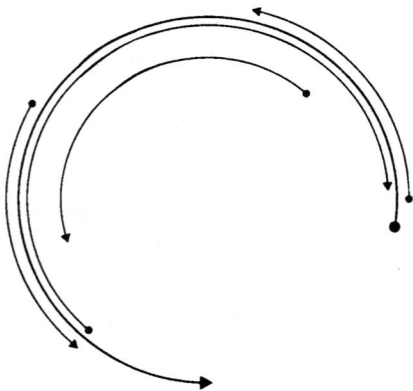

尝试非暴力沟通

Try nonviolent communication

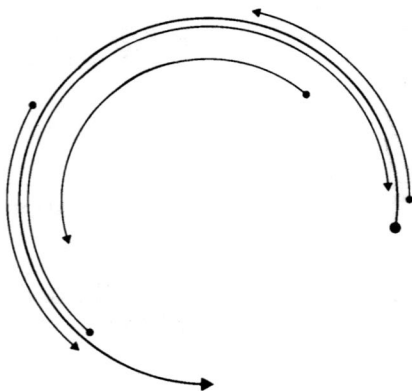

己所不欲，勿施于人

Do unto others as you would have them do unto you

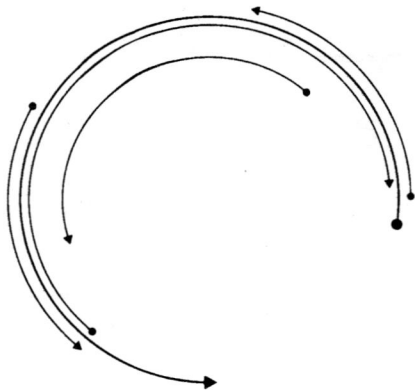

熬夜并不能解决问题

Staying up late doesn't solve the problem

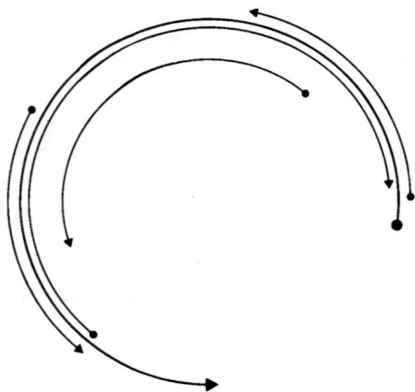

维持原状

Maintain the original state

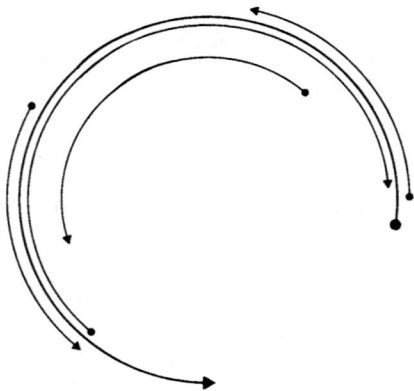

不是所有事都理所应当

Not everyhing can be taken for granted

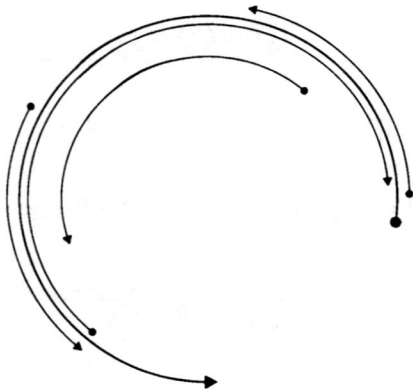

适当让步

Make appropriate concessions

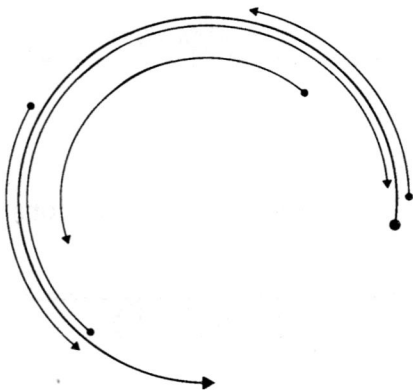

可怕的自律人

What a terrifying self-discipline person

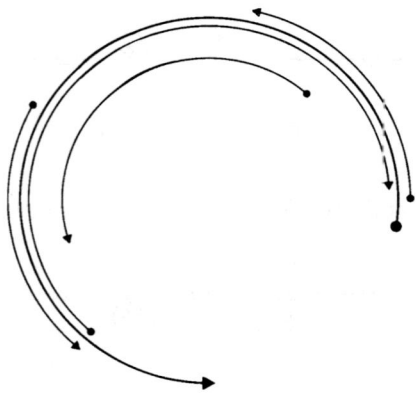

犹豫也是一种选择过程

Hesitation is also a process of selection

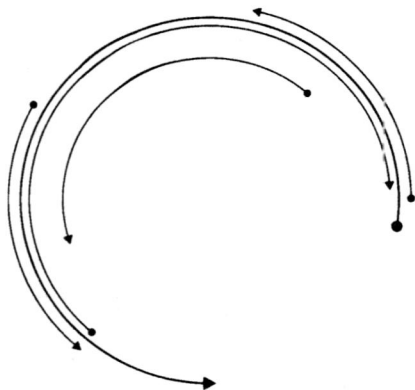

关怀

Show solicitude

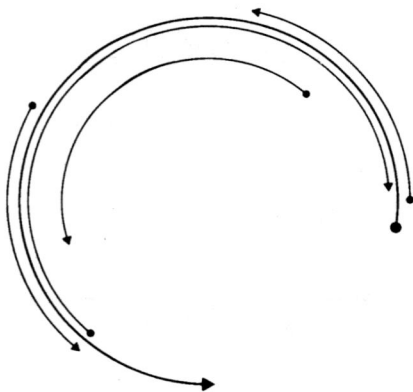

记住自己说过的话

Remember what you have said

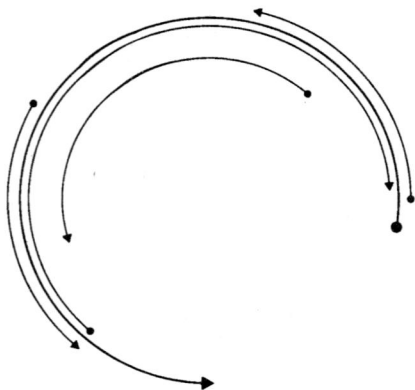

是时候做个了断了

It's time to end it

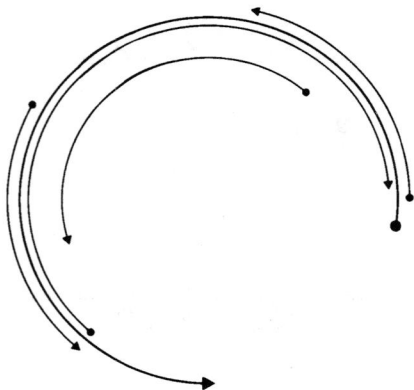

再见

Bye-bye

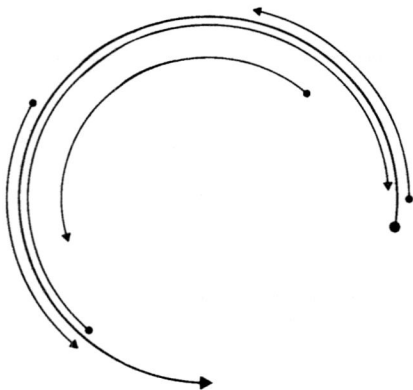

这关乎未来

This is about the future

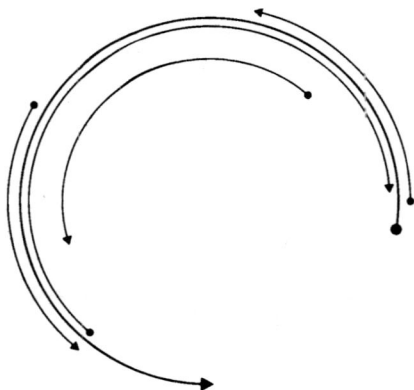

值得去的地方都没有捷径

There are no shortcuts to places worth visiting

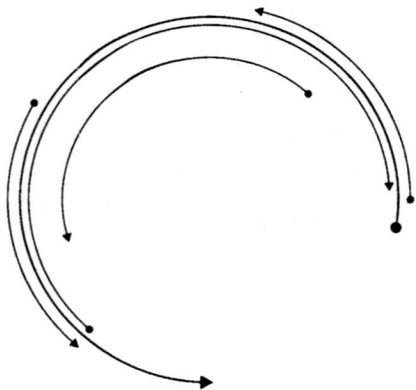

看看身边的人

Look at the people around you

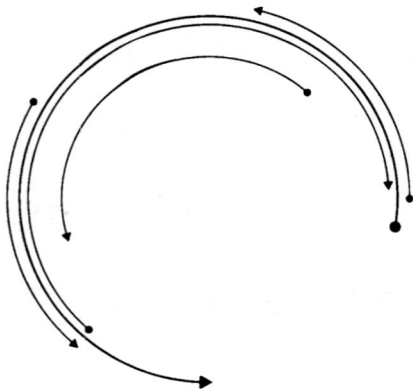

建立亲密关系

Build a intimate relationship

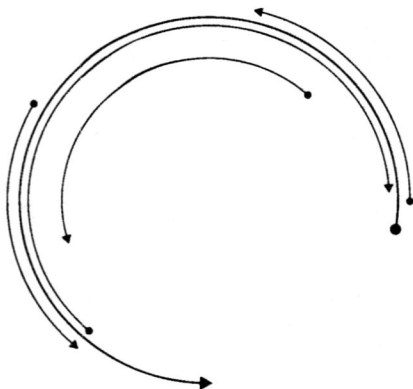

消除顾虑的良方是行动

The best way to eliminate concerns is to take action

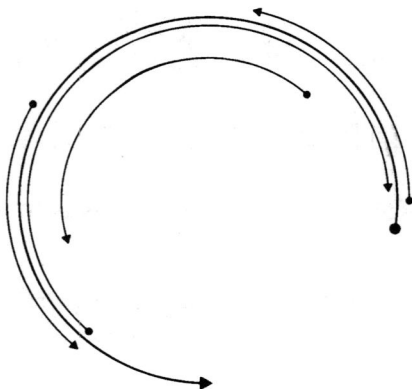

明确表达想法

Express ideas clearly

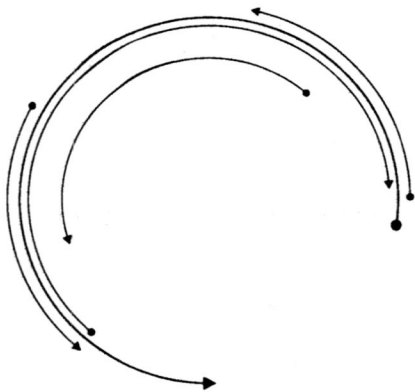

人生没什么不可放下

There is nothing in life that can not be given up

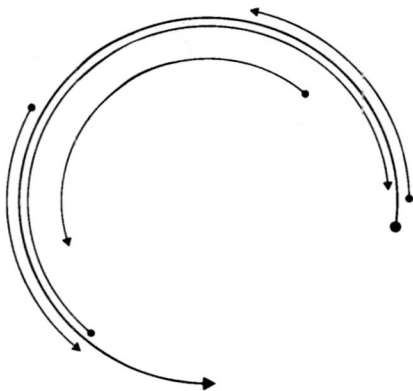

倾听否定，但学会存疑

Listen to negativity but learn to doubt

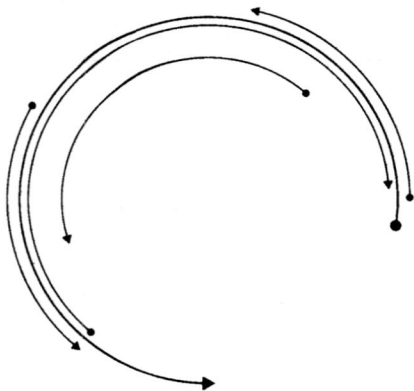

418

忍受痛苦

Endure the pain

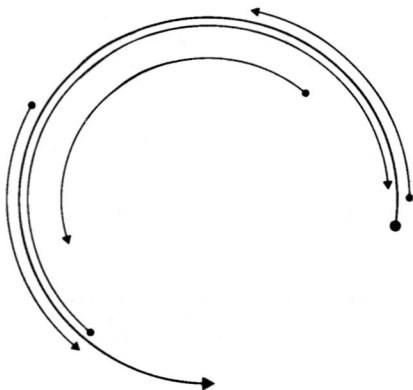

这不是最好的解决方法

This is not the best solution

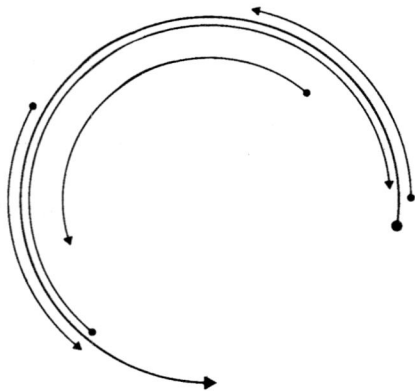

胜利的模样千姿百态

Victory appears in a thousand different poses

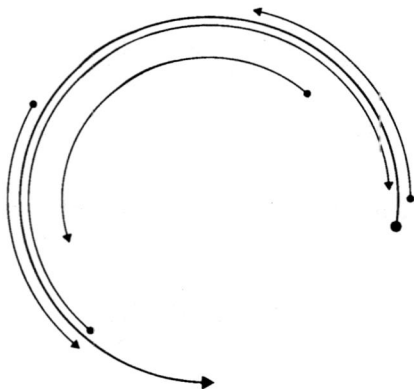

舍与得往往一同出现

Giving and getting often appear together

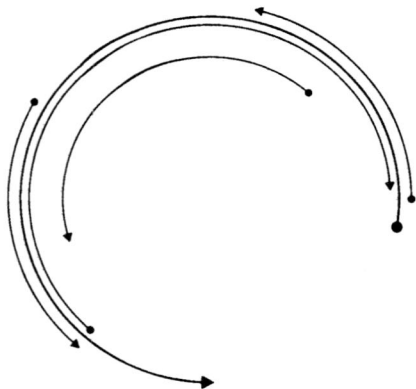

远行

A long journey

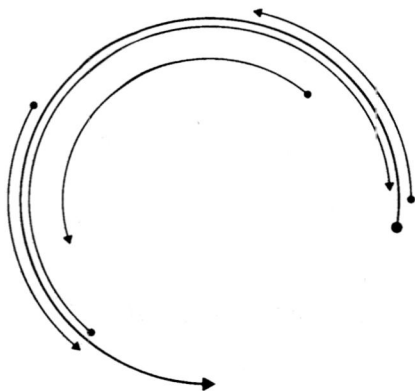

也许是觉醒的好机会

Maybe it's a good opportunity to awaken

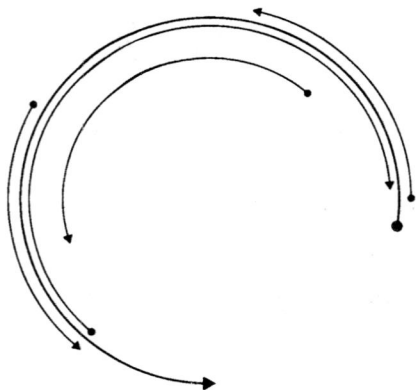

吃点儿好吃的

Eat something delicious

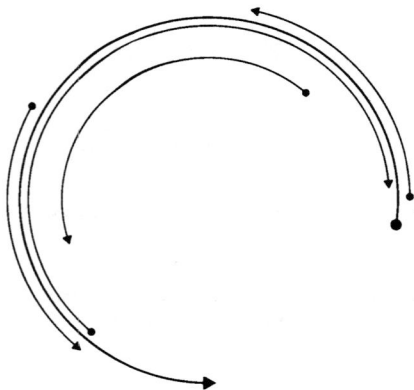

你会发现这件事需要自己独立完成

You will find that you need to do this on your own

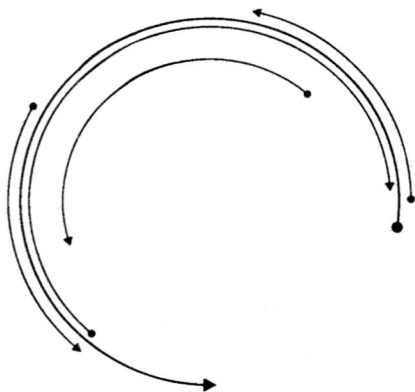

拥抱

Hug

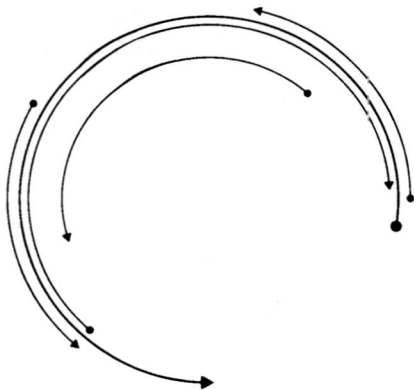

追求有意义的

Pursuing meaningful

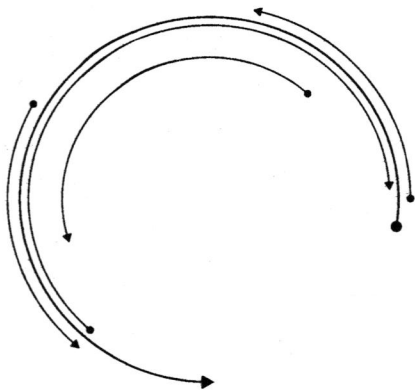

尝试接受

Try to accept

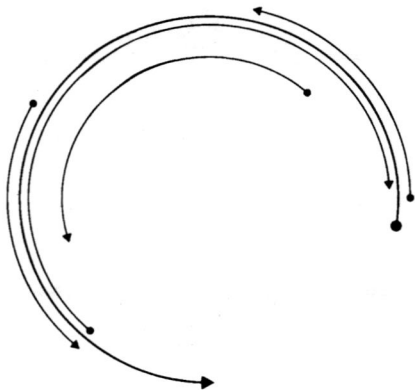

坚持，但不要固执

Insist but don't be stubborn

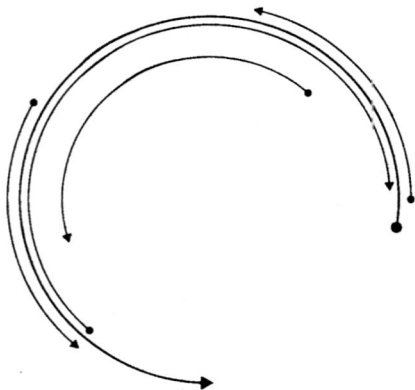

记录偶尔的灵光

Record the occasional inspiration

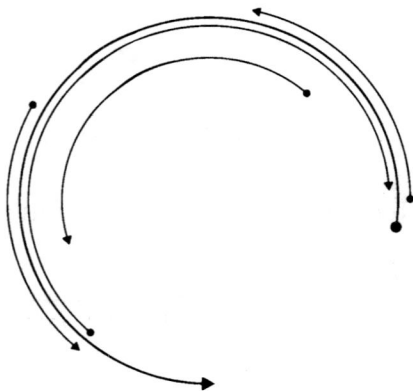

失望

Disappointed

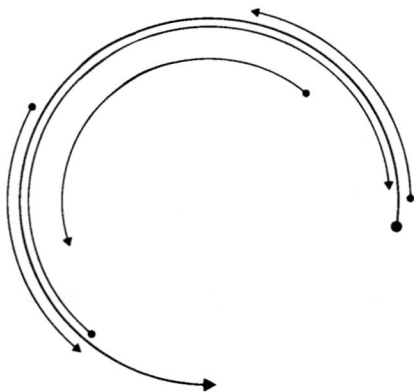

学会向内寻求力量

Learn to seek power within yourself

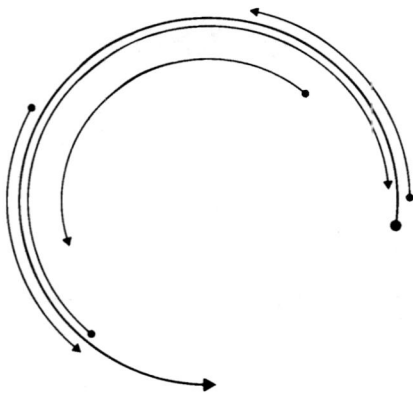

刚刚好

Just right

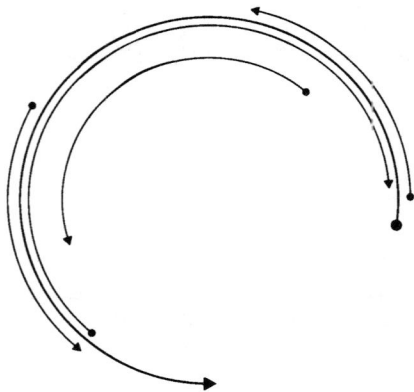

你很厉害

You're awesome

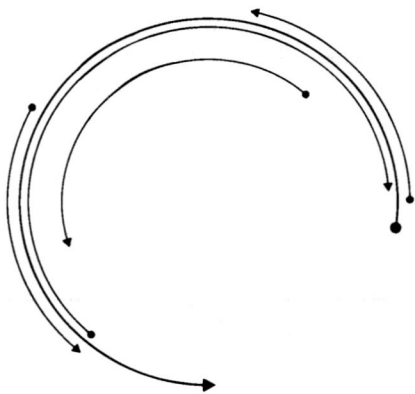

实践它

Practice it

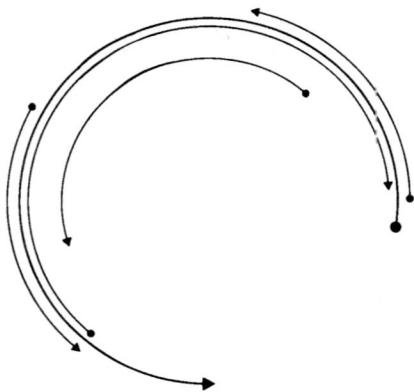

列出可行性与阻力

List feasibility and resistance

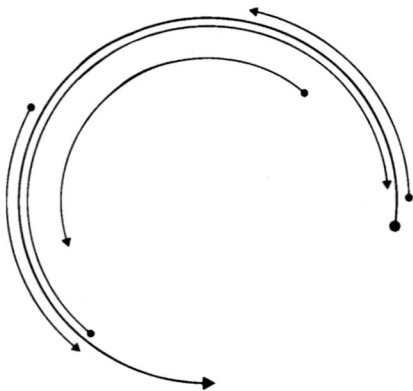

理智规划

Planning rationally

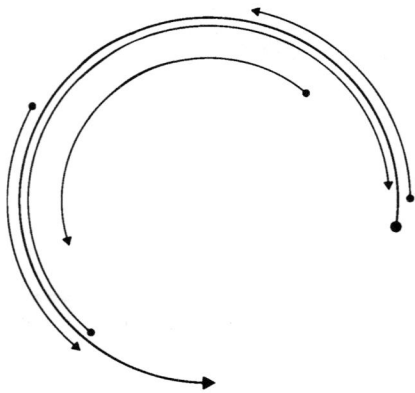

458

故人

Old Friend

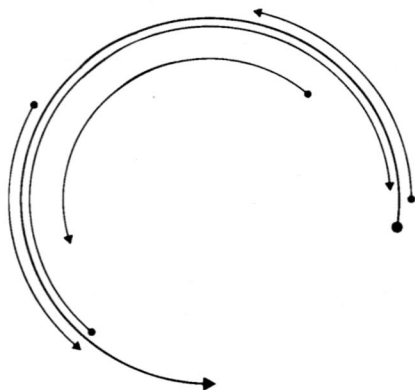

你需要清除自身障碍

You need to remove your own obstacles

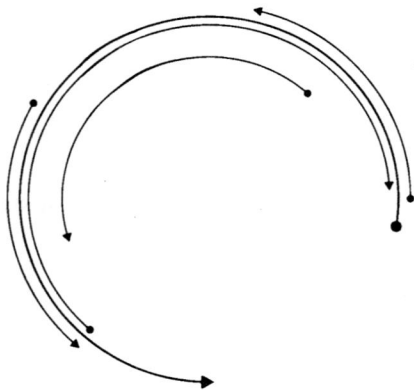

不要明知故犯

Don't knowingly break the rules

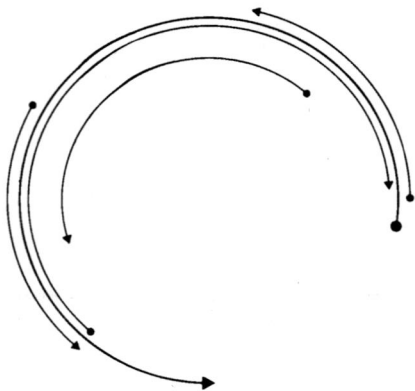

尝试另一种解决方式

Try another way to solve the problem

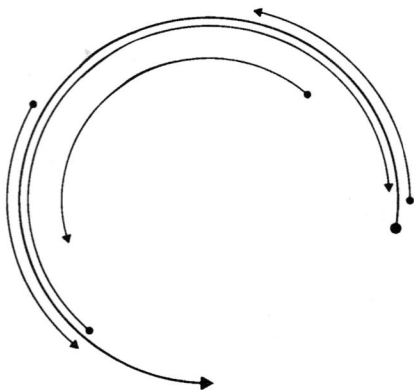

再慷慨一些

Be more generous

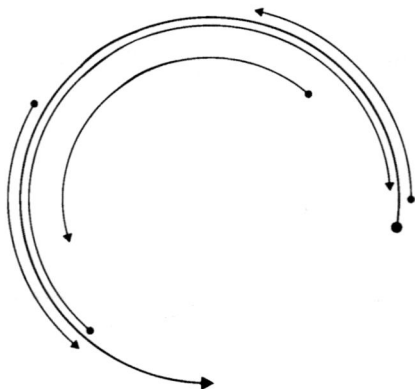

原谅

Forgive

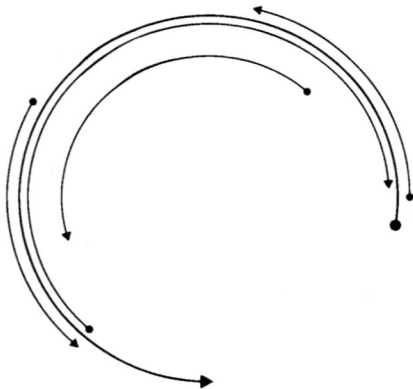

会有人理解你的

Someone will understand you

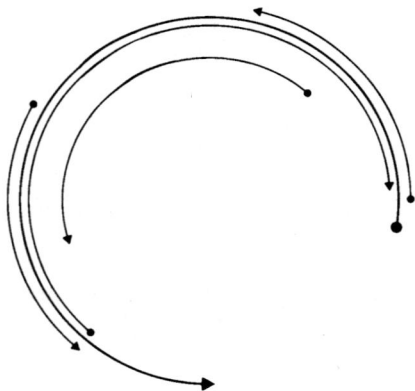

静观其变

Wait and see what happens

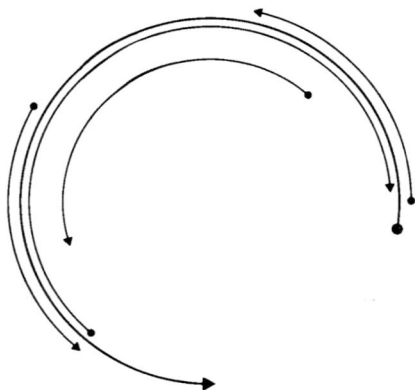

现在就是最好的时机

There's no better time like the present

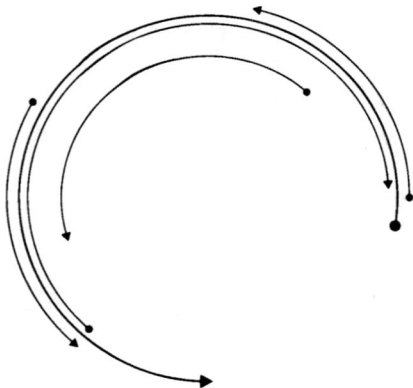

476

你当像鸟飞往你的山

Fly as a bird to your mountain

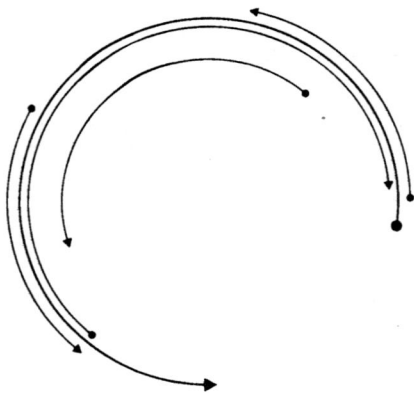

捍卫

Defend it

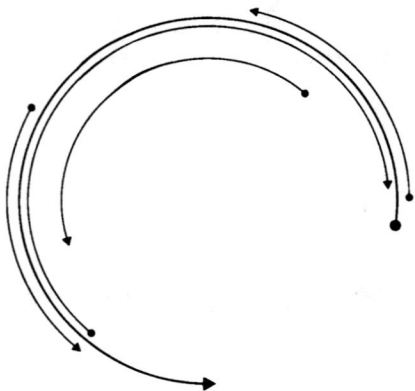

不值得去争取

It's not worth fighting for it

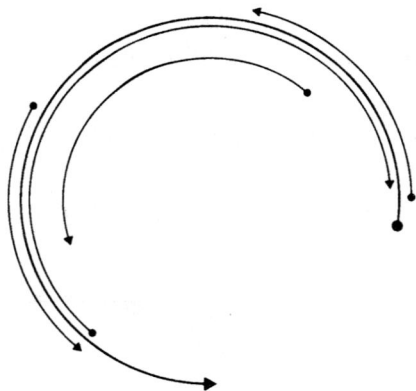

尝试不太可能的解决方案

Try an unlikely solution

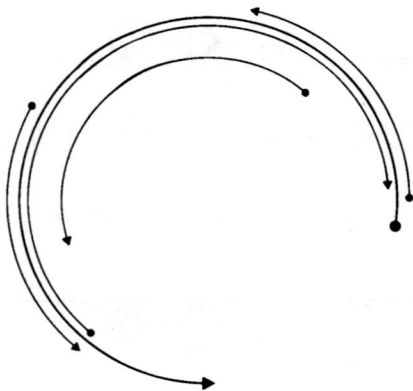

好好告别

Say good-bye

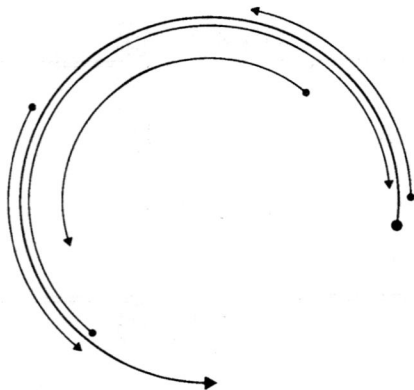

486

你失去的会以其他方式归还

What you lost could be retuned in another way

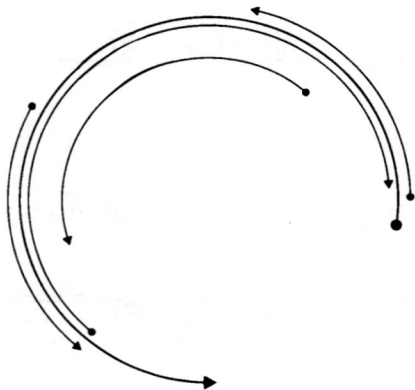

背叛

Betray

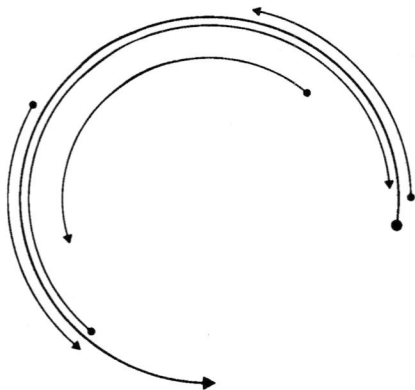

让复杂的事变简单

Simplify complex things

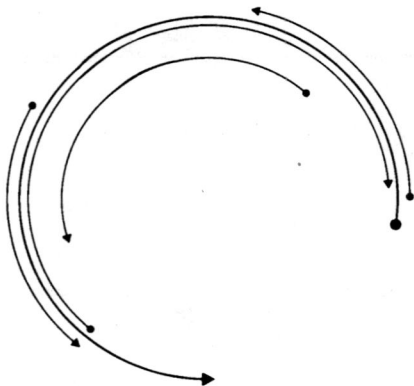

492

特立独行

Maverick

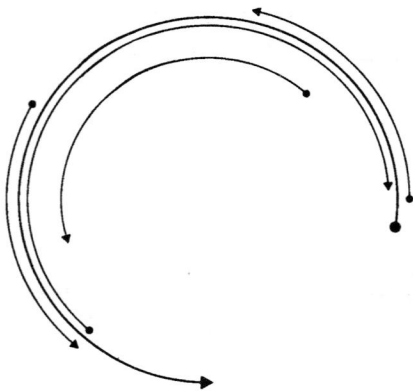

不求回报的付出

Giving without expecting anything in return

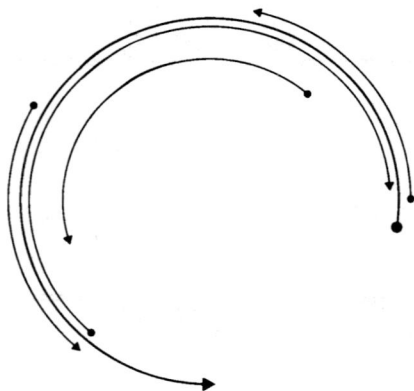

一个人的长路

The long road of one

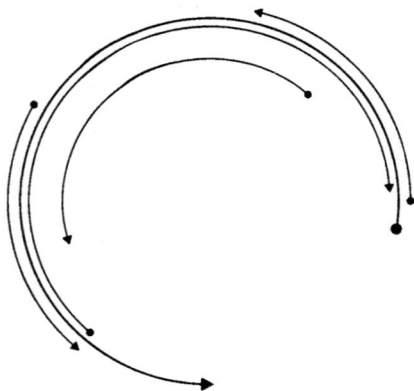

出发

Let's go

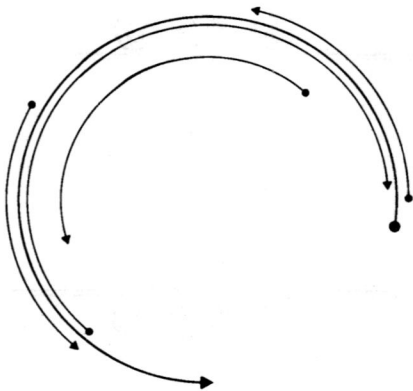

其实你知道

Actually, you know it

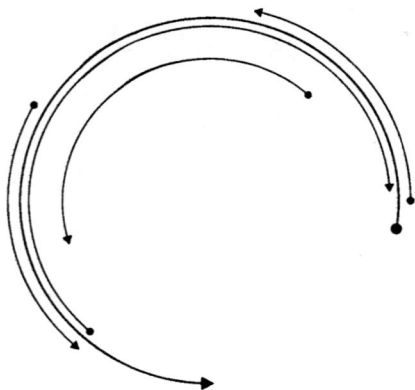

可能不会得到认同

Probably won't get approval

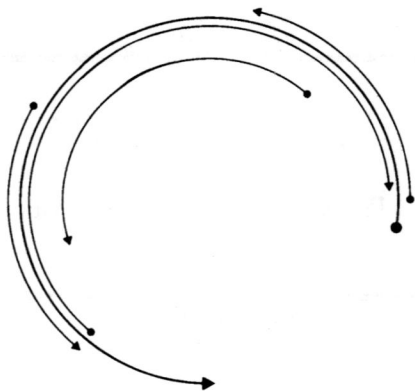

往前走，别回头

Go forward and don't look back

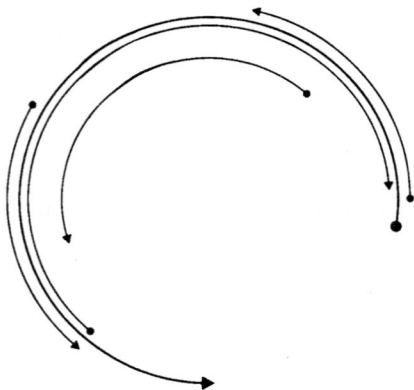

尽你所能地开心

Do your best in being happy

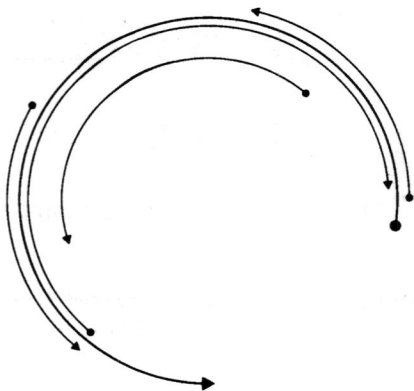

完美

Perfect

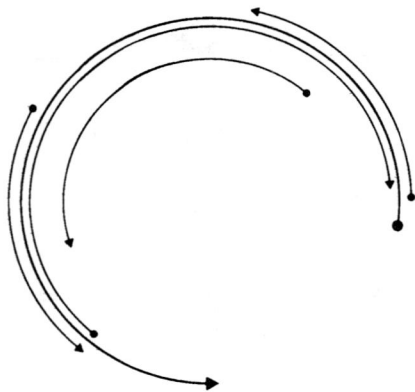

无意义的让步

Meaningless concession

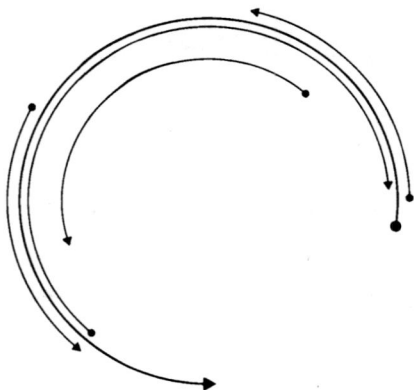

泾渭分明

Quite different from each other

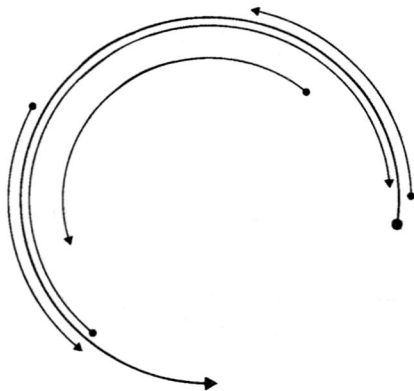

这不言而喻

It goes without saying

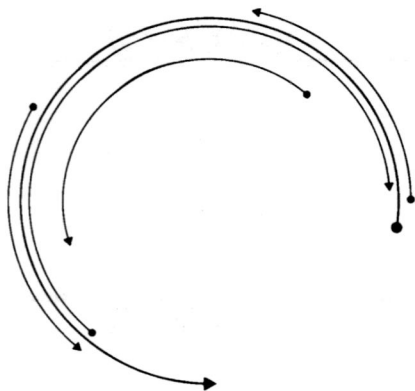

一场梦

Just a dream

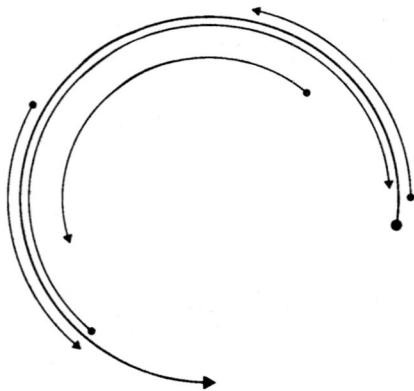

悲伤是不可避免的

You can not keep away from grief

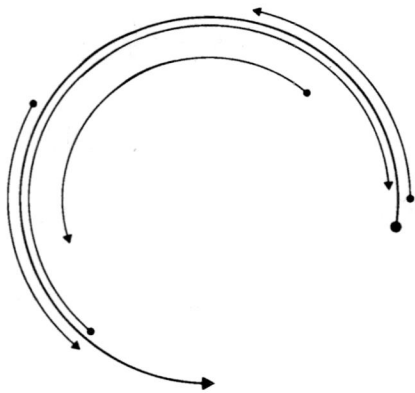

值得庆祝

Worth celebrating

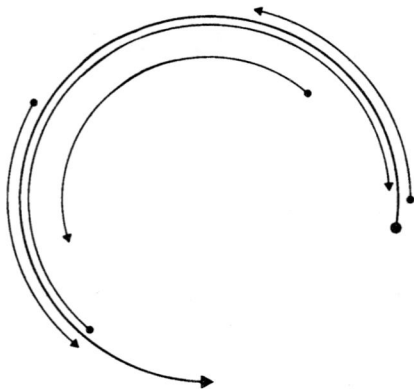

最好的解决方法未必显而易见

The best solution is not necessarily obvious

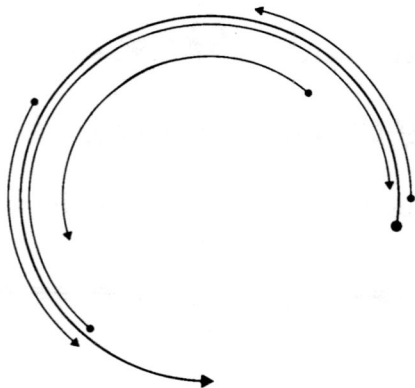

总会留有遗憾

There are always regrets

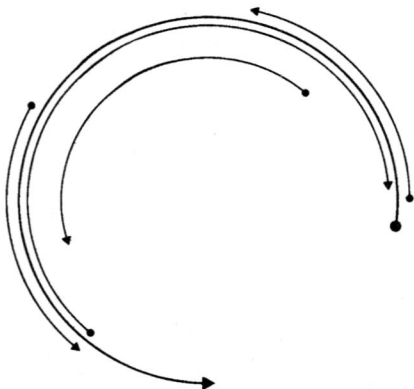

可能会浪费金钱

May be a waste of money

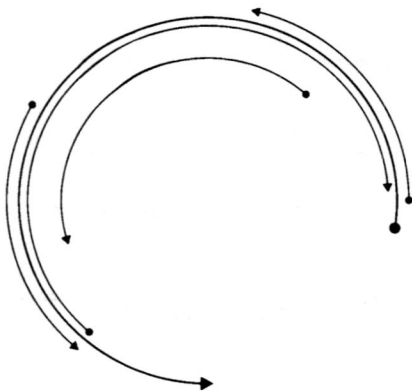

融洽

Harmonious

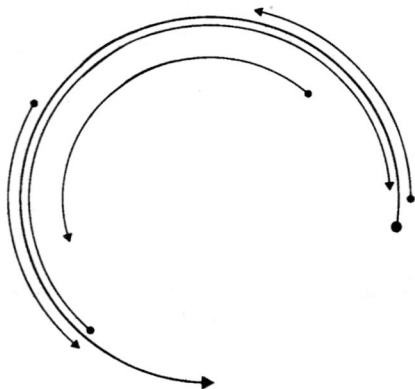

胸有成竹

Have a well-thought-out plan

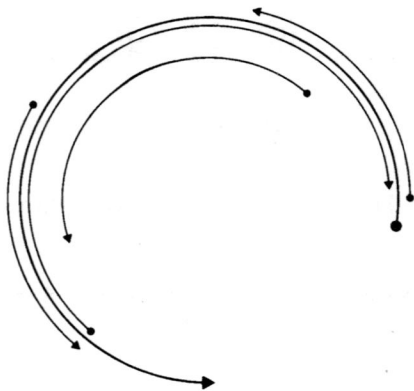

守候

Wait

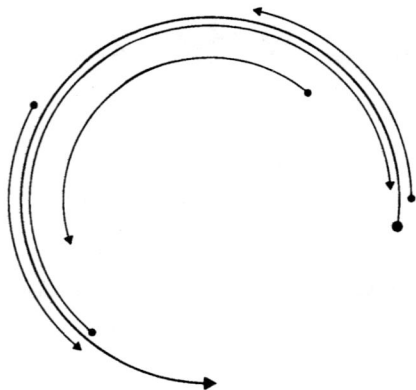

这太糟糕了

That's too bad

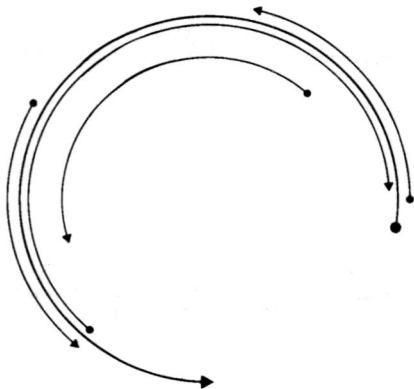

眼见非实

Seeing is not believing

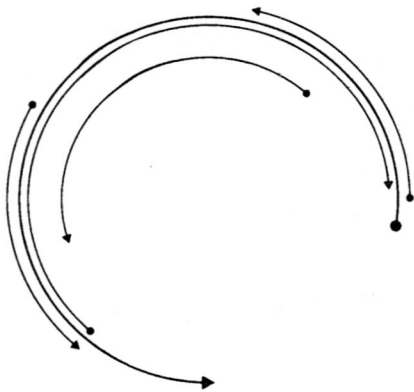

538

明天会是新的一天

Tomorrow is another day

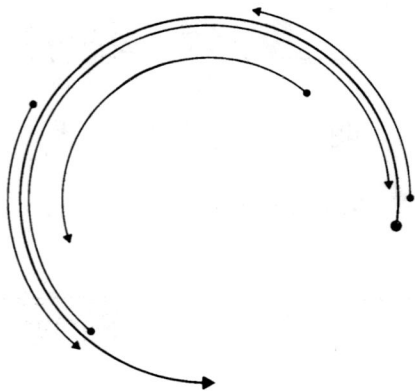

试着大哭出来

Try to cry out loud

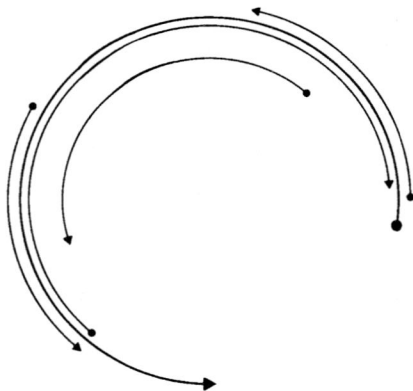

独处

Stay alone

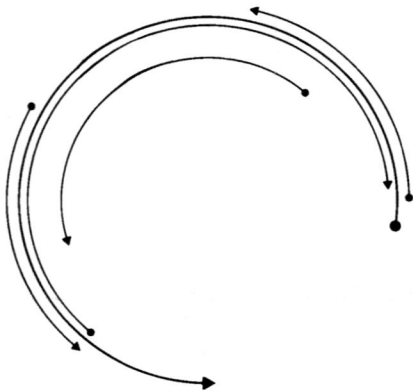

慎行

Proceed with caution

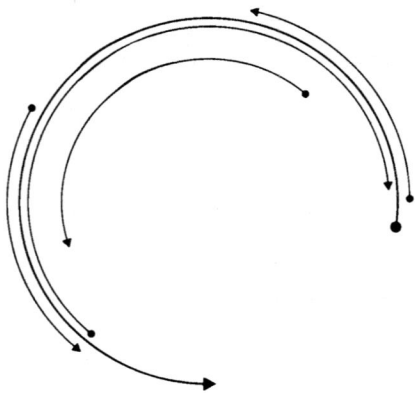

一无所有

Have nothing at all

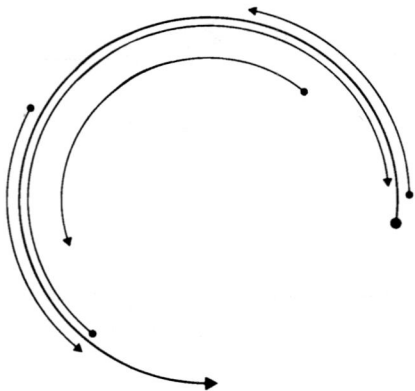

放过

Let it pass

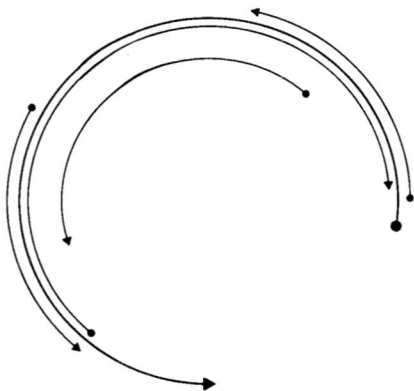

停下来

Stop it

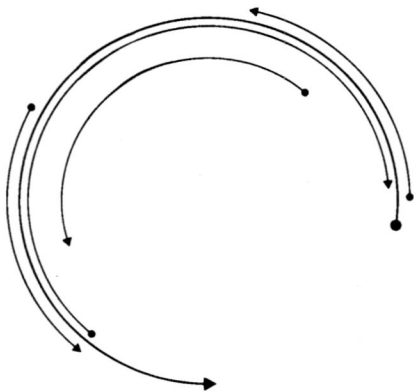

保守秘密

Keep the secret

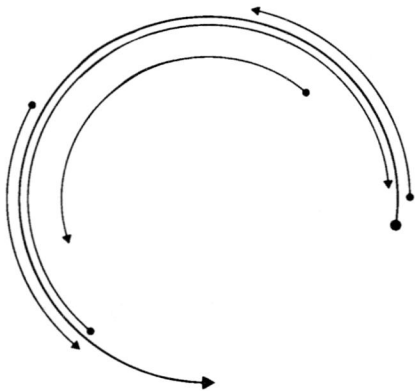

存在一定困难

There are some difficulties

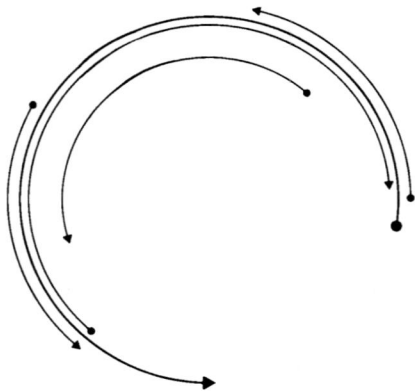

很难愈合

It's hard to heal

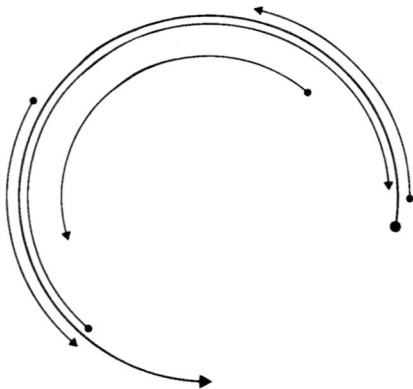

一往无前

Press forward with indomitable will

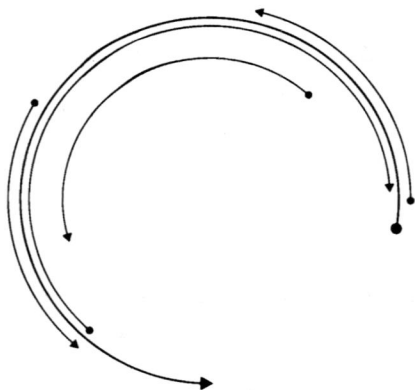

悲伤会淡去

Sadness will fade away

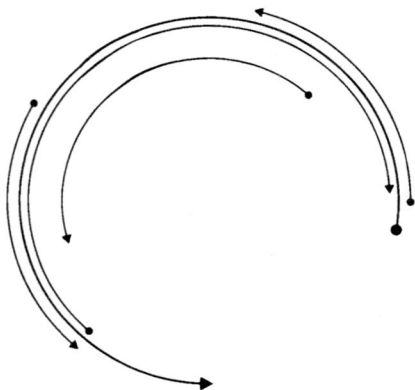

叫醒沉睡的自己

Wake yourself up

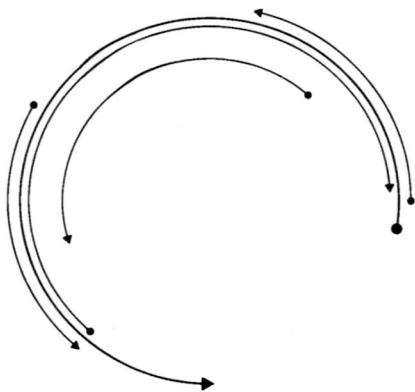

空想家

A dreamer

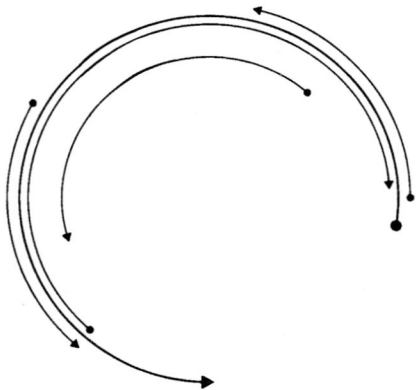

活力

Vitality

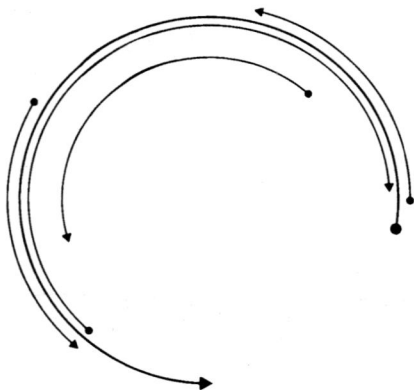

重获新生

A new lease of life

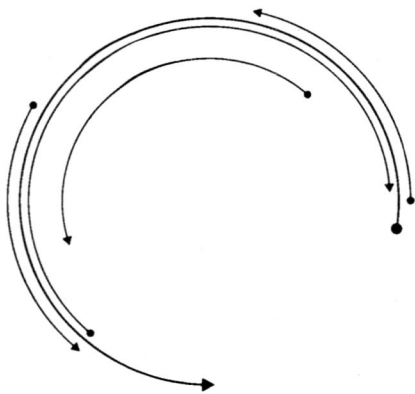

你正在被消耗

You are being sabotaged

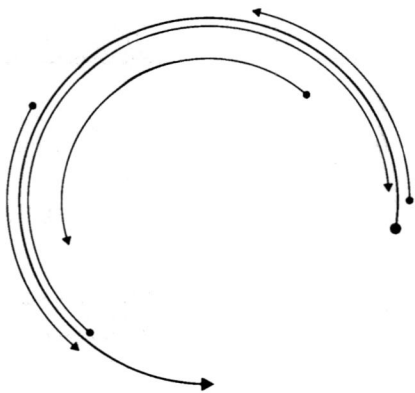

没关系

It's OK

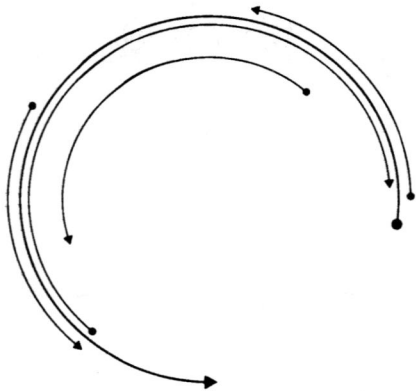

574

这是多余的

It's unnecessary

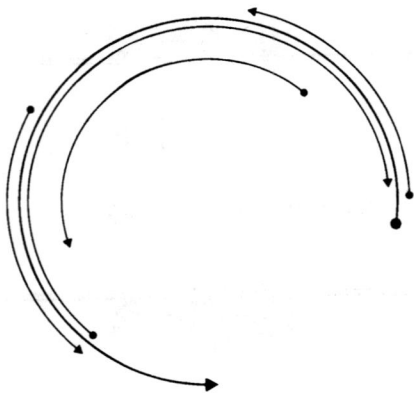

精进

Be absorbed and desirous to do better

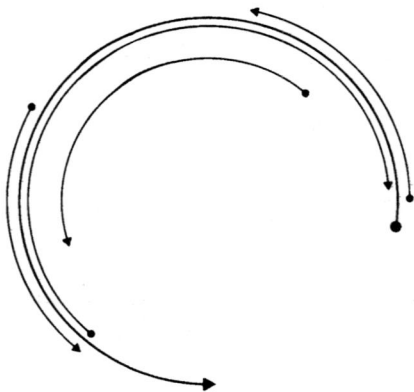

没什么是绝对的

Nothing is absolute

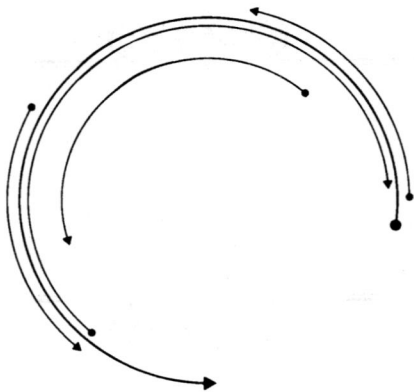

好的预感

A good feeling

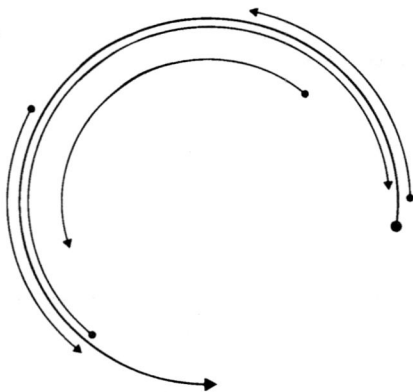

回馈

Feedback

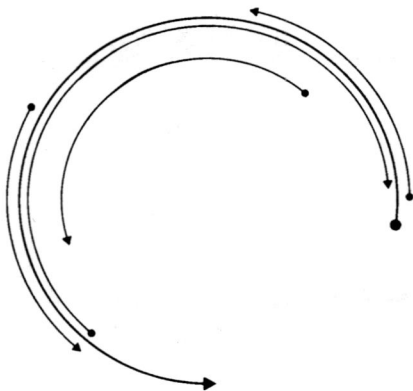

保持良好心态

Keep a good mood

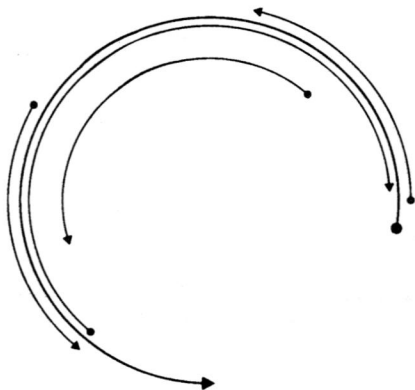

天会放晴的

It will be sunny

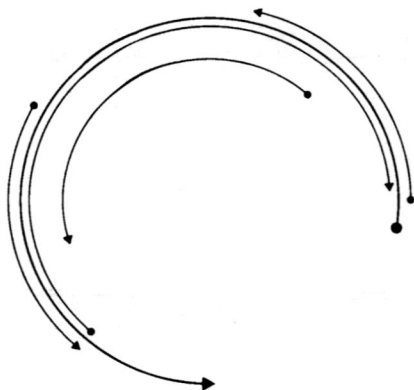

第六感

The Sixth Sense

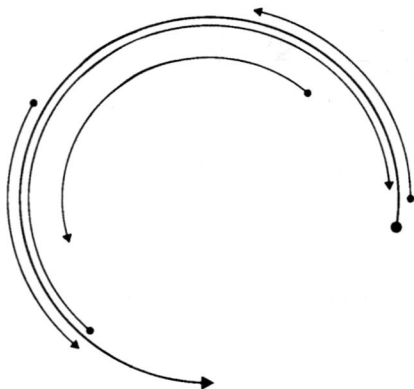

釋怀

Make peace with yourself

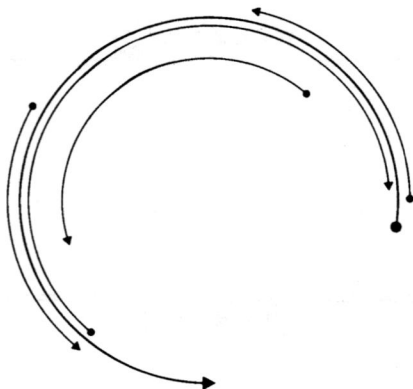

咫尺天涯

So near but yet so far

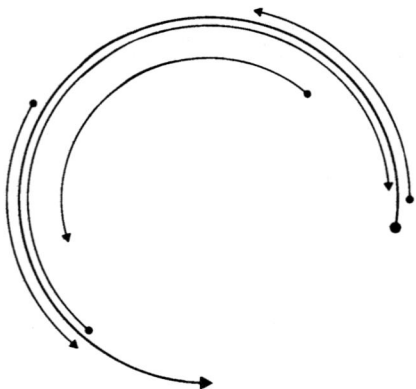

契机

Opportunity

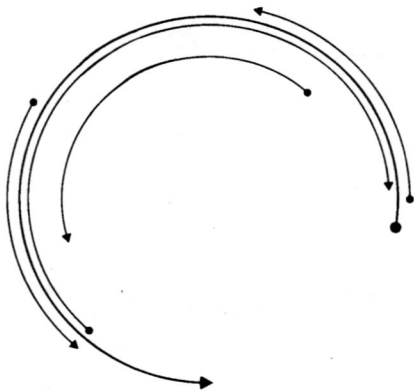

不要轻易判断

Don't judge easily

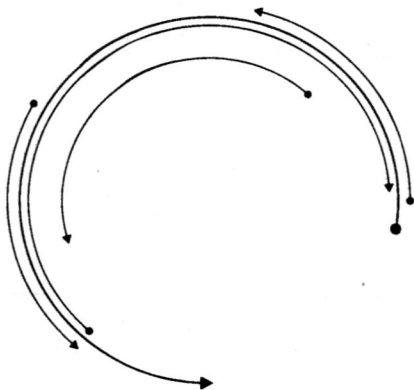

控制

Control it

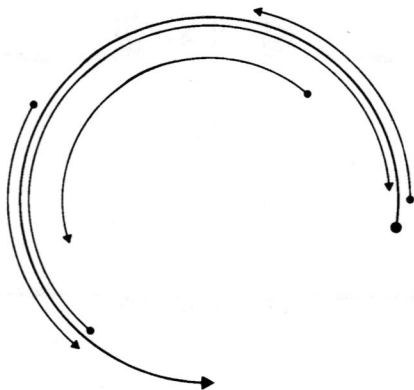

适可而止

Enough is enough

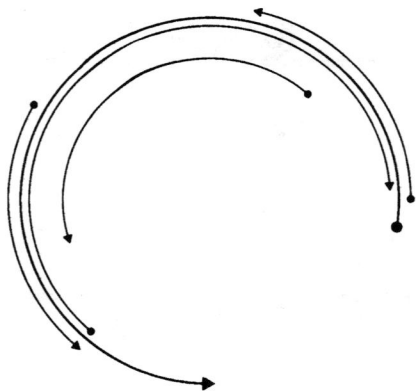

不要制造麻烦

Don't make trouble

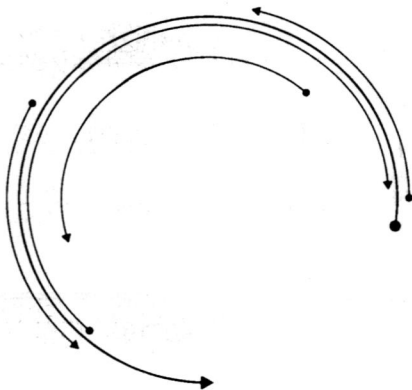

漫长的等待

A long wait

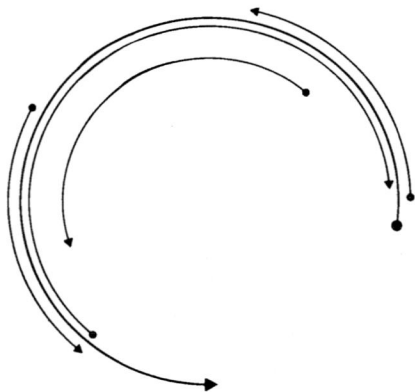

606

傻一点没坏处

It's not bad to be silly

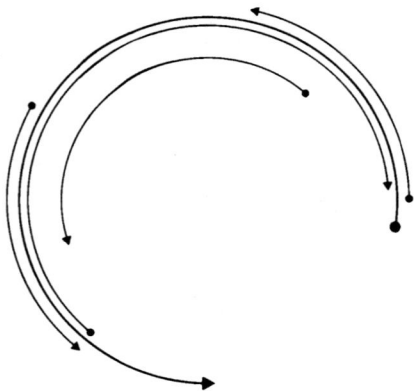

608

追不回的昨天

You can't go back to yesterday

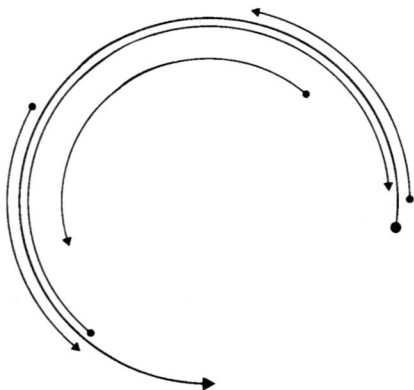

南辕北辙

Poles apart

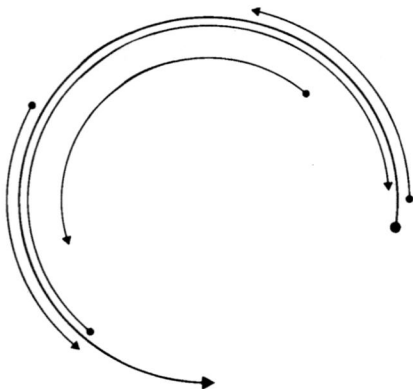

何必介怀

Why bother?

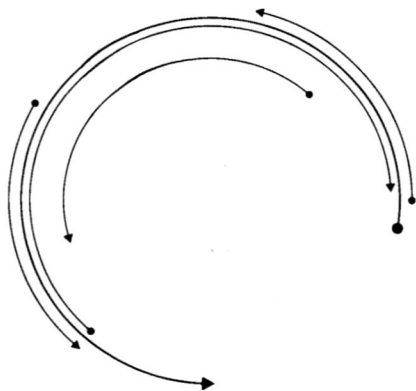

试着面对最真实的想法

Try to face your true thought

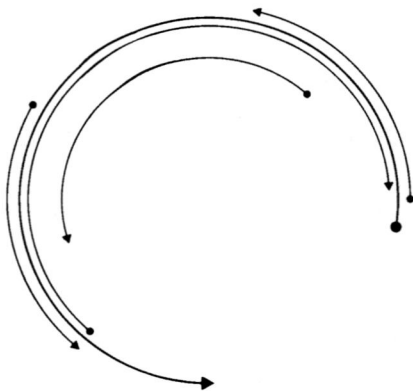

一场未知的旅程

An unknown journey

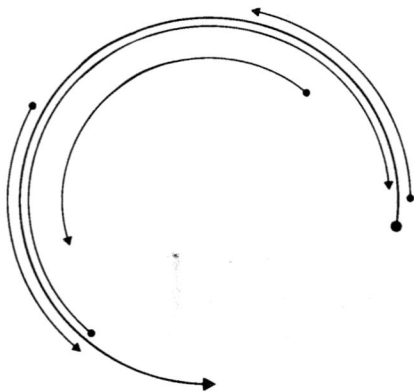

孤岛

An isolated island

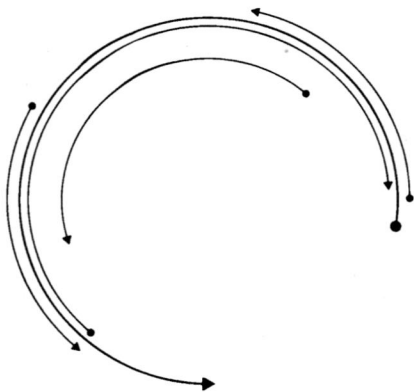

心渐渐冷掉

Gradually become dispirited

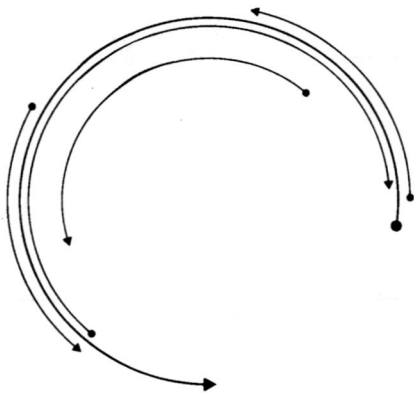

忍耐

Bear with it

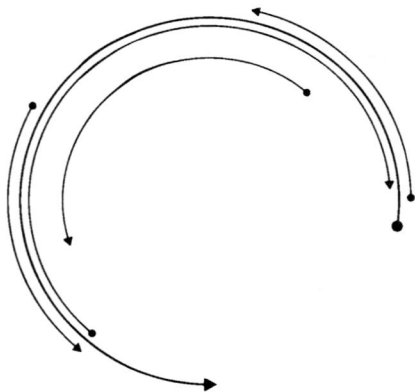

有些人选择了其他道路

Some people took another path of life

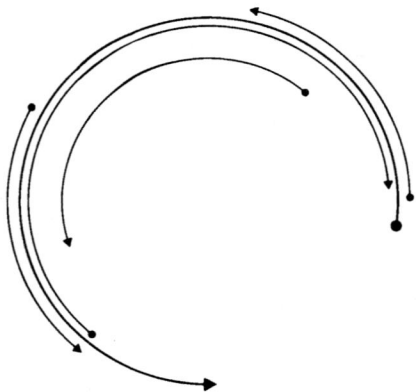

会给身边人带去温暖

It will bring warmness to people around you

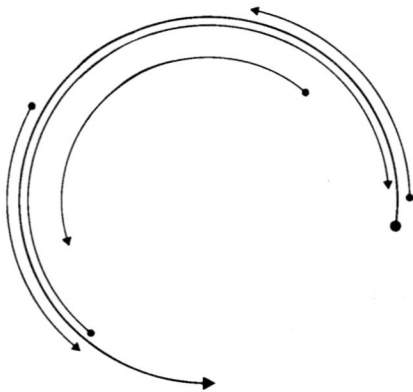

628

明媚

Bright and beautiful

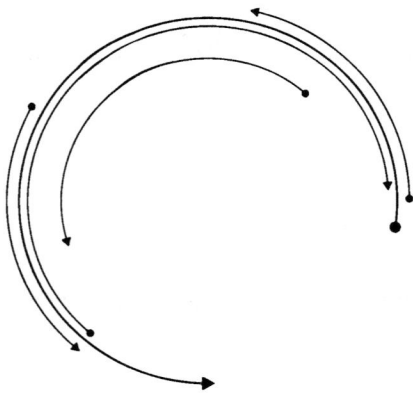

美好的

Wonderful

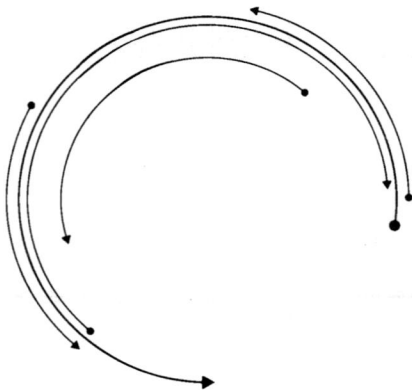

容许不确定性

Allow uncertainty to exist

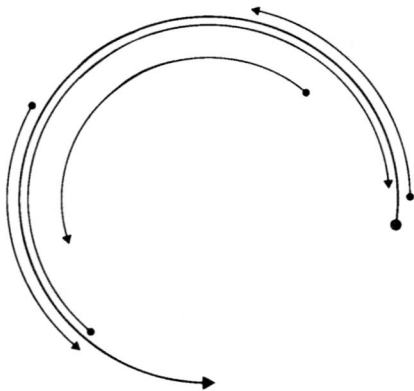

634

好与坏都是一种结果

Both good and bad are the same outame

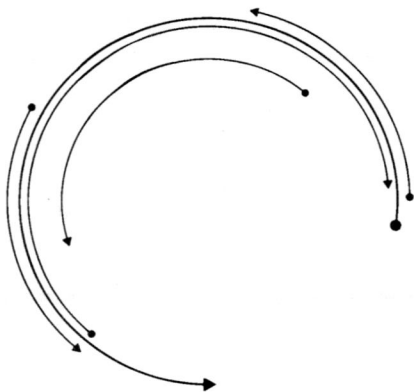

636

投桃报李

Return a favor with a favor

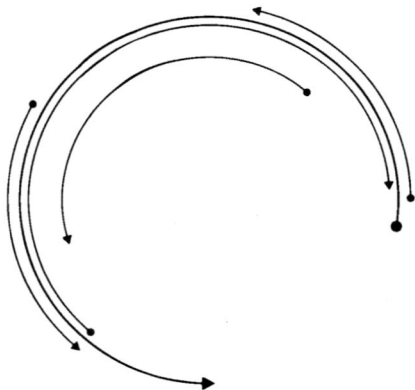

拥有被讨厌的勇气

Have the courage to be hated

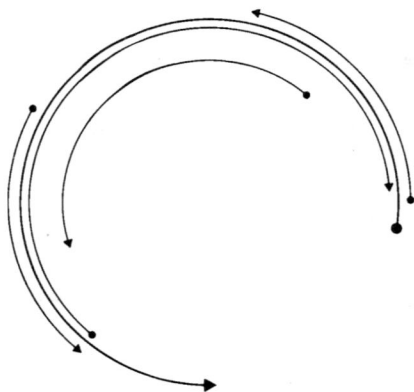

640

与时间做朋友

Making friends with time

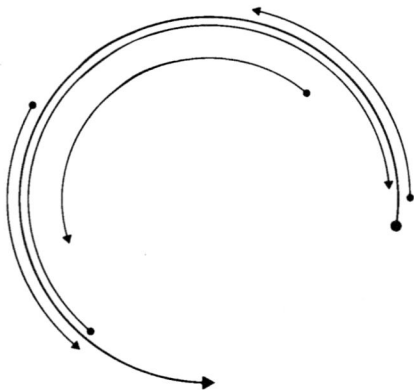

养成好习惯

Develop good habits

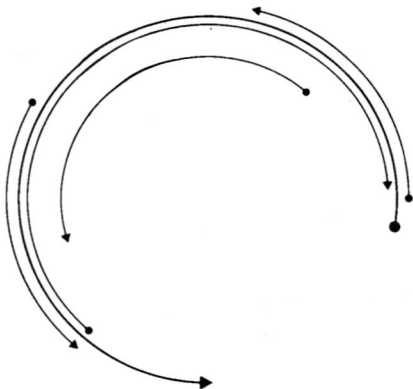

644

钝感力

The Power of Insensitivity

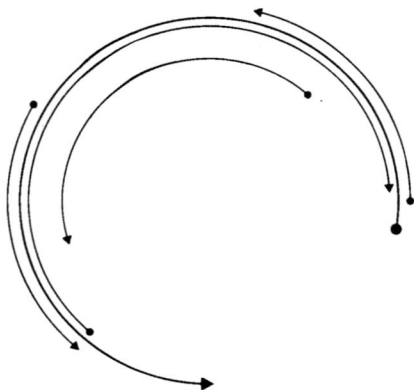

扔掉一些东西

Get rid of something

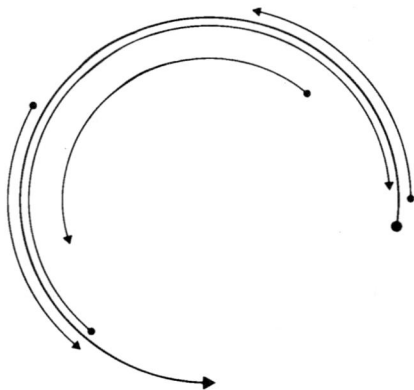

不要轻视他人

Don't look down upon others easily

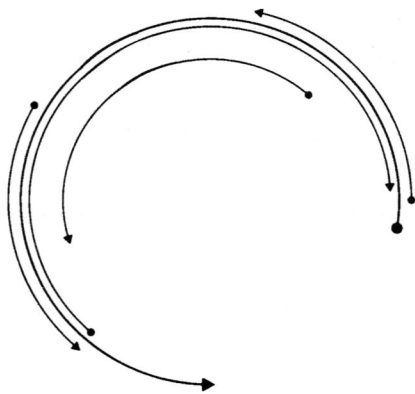

650

变通也是一种选择

Being flexible is also an option

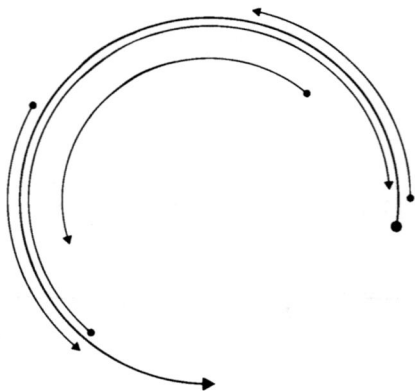

简单

Easy

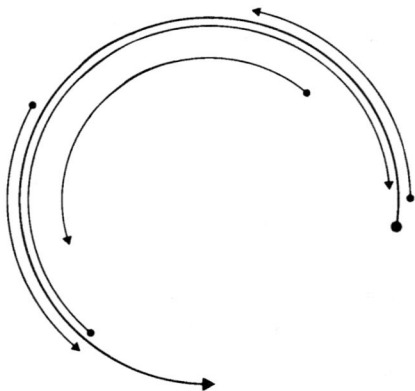

像白乌鸦一样罕见

Rare as the white crow

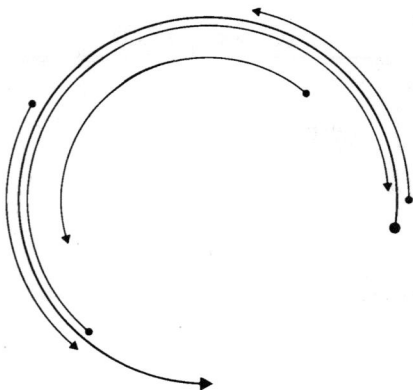

656